SUPERPIGS
AND
WONDERCORN

SUPERPIGS

AND

WONDERCORN

THE BRAVE NEW WORLD OF BIOTECHNOLOGY AND WHERE IT ALL MAY LEAD

Dr. MICHAEL W. FOX

Lyons & Burford, Publishers

Library of Congress Cataloging-in-Publication Data

Fox, Michael W., 1937-
    Superpigs and wondercorn: the brave new world of biotechnology
and where it all may lead/Michael W. Fox.
      p.   cm.
    Includes bibliographical references and index.
    ISBN 1-55821-182-9
    1. Genetic engineering—Social aspects. 2. Genetic
engineering—Environmental aspects. 3. Biotechnology—Social
aspects. 4. Biotechnology—Environmental aspects. I. Title.
TP248.6F69   1992
174'.9674—dc20   92-18003
                                                 CIP

FOR
Deanna

Those who would take over the Earth and manage it
I see they cannot grasp it;
for the Earth is a spiritual vessel
and cannot be forced.
Whoever forces it spoils it.
Whoever grasps it loses it.

—LAO TZU, *Tao Te-Ching*

Contents

ACKNOWLEDGMENTS

I wish to express my gratitude to my wife Deanna Krantz for her support, critical discussion, and review of my work while preparing this book, and also to my executive assistant Ellen Truong for typing many drafts. I am also indebted to my publisher Peter Burford, whose editorial hand did much to improve my original manuscript.

Preface

The world is entering a new scientific age: the genetic age of biotechnology. Genetic engineers have broken the genetic code, just as in the 1930s and 1940s physicists broke the code of the atom. Now they are developing ways to alter the genetic structure of living things by removing certain genes and inserting the genes of one species into another.

Genetic engineering allows living things to be made into "new" and ever more useful and profitable commodities. Earlier industrial epochs transformed various natural resources into new forms of energy and synthetic materials. But the genetic age has more profound consequences. Although this new age of biotechnology promises many benefits, it also raises serious risks and fundamental questions that need to be considered by all, not just policymakers or academicians but the public at large. The genetic age is now upon us. For better or worse, it will affect our lives and those of all generations to come.

INTRODUCTION

*T*he genetic code was broken with the discovery of the structure of DNA, the molecule that carries the genetic blueprint of every living organism. Various techniques have been developed, and some patented, for the identification, splicing, and commercial use of certain genes in animals, plants, and microorganisms. With microscopic aids, researchers can inject genetic DNA material into the cells of plants and animals. Human genes that regulate growth or govern the manufacture of insulin or the production of other biologics, such as interferon, have been isolated and "spliced" into the genetic code sequences of bacteria. Fermentation—cultures of such bacteria—now enables the mass production of human insulin for diabetics and other products of use in medicine and industry.

A diversity of other genetic engineering techniques have been developed in many different species, just a few of which will be detailed here. Monoclonal antibodies, of great diagnostic value in medicine, are manufactured using the mutant "nude" mouse as living factories. Transgenic animals—creatures that have others' genes spliced into their genetic code—especially mice, are also being used increasingly in toxicology and drug evaluation.

A weakened form of the avian leukosis virus has been inserted into poultry and passed on to three generations of birds. It is planned to use this virus as a carrier for inserting genes of desirable traits such as disease resistance or rapid growth.

The defective genes of certain heritable disorders in humans, such as neurofibromatosis, have been spliced into the cells of embryo mice to create "models" of the disease. Viruses, like the AIDS virus, have likewise been spliced into mouse embryo cells to make a line of AIDS-carrying mice. Mice with inherited diseases that cause sterility and neurologic problems have been cured by inserting normal mouse genes into defective embryos. These "germ-line remedies" are too hazardous yet for human application. But somatic cell remedies, such as injecting genetic material into bone marrow, hold great promise.

Human genes that are involved in the production of such biologics as Factor IX, which is deficient in some hemophiliacs, have been put into sheep embryos. In later life the sheep secrete these substances in their milk. The genetic engineering industry sees this as the transformation of animals into "protein factories" or "bioreactors," sometimes even referring to them as "pharm animals."

The genes of bacteria that kill insect pests are being developed in agriculture to protect the roots, or sometimes the whole plant, from pests. For instance, a tomato plant with an inserted bacterial gene produces a toxin that makes the plant resistant to caterpillars. Plants may soon produce spider venom to kill insect pests and develop fish antifreeze genes to resist frost.

Other bacteria that produce poisons to kill plant pests are being developed as purportedly safer alternatives to chemical pesticides. A toxin-producing gene from one strain of bacteria has been spliced into another related bacterium that lives in the roots of corn plants to kill root-invading pests. And plants are being engineered to be resistant to herbicides and even to produce their own fertilizers and chemical insecticides.

We know nothing of the potential environmental risks and long-term consequences of any of these activities. The complex ecology of Earth may soon be transformed into an industrial "biosystem," with potentially catastrophic environmental consequences. However, *this new technology could be used appropriately*, as, for example, to develop microorganisms to synthesize essential biologics, such as insulin and antibodies; to facilitate the recovery of various ores and minerals; to treat water and sewage by removing or detoxifying pesticides, heavy metals, and other industrial and agrichemical poisons; and to engineer plants to help halt the spread of deserts.

I am not in principle opposed to biotechnology. It is not an unexpected development in the evolution of civilization. However, I am opposed to those who seek to use it only for profit regardless of ethical, social, environmental, and animal-welfare concerns. If the twentieth century has taught us anything it is that the Earth is too precious to be plundered, and that we do so at our own risk.

I believe we need to move away from the pose of our industrialist forefathers, who saw nature and all living things less as kin than as resources to exploit, devoid of inherent value or mystery. A new postindustrial mind-set suited for the awesome task of developing the biotechnology industry so that it is applied appropriately, even *creatively* and *profitably*, must break with the orthodoxy of a consumer society. We must, for example, break away from our addiction to meat and not misuse this technology to support and even expand an ecologically unsound meat-based agriculture. We can use this technology creatively to repair this despoiled planet and to help "dress and keep" the Garden of Eden. More land can be returned to nature only if we shift away from a meat-based agriculture to a regenerative, sustainable eco-agriculture where the world production and consumption of animal protein (including marine sources) decline and the production and consumption of vegetable protein increase. Biotechnology could be of great value in this kind of agriculture by developing new varieties of more nutritious, more easily harvested crops that are more resistant to local diseases and climatic variables. To genetically engineer seeds resistant to herbicides and to spray crops with bacterial pesticides is not a creative use of biotechnology.

Biotechnology's appropriate application in medicine could do much to alleviate human suffering and disease. But real advances in human medicine are not likely to come until the health industry recognizes that a medical science that relies primarily upon crude and often cruel laboratory experiments on animals for its advancement is conceptually flawed and ethically blind, because no good can come, in the long term, from laboratory animal vivisection.[1]

---

[1] Many altruistic vivisectors claim that all medical advances of any significance have come through experimenting on animals. This is scientifically invalid; most advances have come through improvements in nutrition, sanitation, and preventive medicine. See M. W. Fox, *Inhumane Society: The American Way of Exploiting Animals* (New York: St. Martin's Press, 1990).

It is for these reasons that I am opposed to the utilitarian application of genetic engineering to animals. It could be applied appropriately to help correct genetic defects and facilitate the preservation of rare and endangered species, but a new attitude toward animals and nature, indeed toward the whole of creation and the creative process itself, is needed if this technology is to be appropriately applied. If it is applied in the old materialistic mind-set, then the end of the natural world will be our legacy. This book raises more questions than easy solutions, but the very act of asking these questions may ensure that our children and the rest of creation will not suffer the legacy of a world impoverished by genetic engineering run amok.

CHAPTER 1

BIOTECHNOLOGY: THE WORLD MADE OVER

Our view of nature will influence the way we treat nature, and our view of human nature will affect our understanding of human responsibility.
—IAN BARBOUR, *RELIGION IN AN AGE OF SCIENCE*

*O*ver the past decade and a half, the biotechnology industry has become the fastest growing industry in recorded history. Venture capitalists have invested in both small and large companies that promise great profits from new, genetically engineered products and services. By 1984, investment in genetic engineering had already exceeded $2.5 billion, with the total global investment estimated at $40 billion by the year 2000, according to analyst

J. Elkington.[1] Sales of biotechnology products in the United States reached $1.11 billion in 1989 and are expected to surpass $40 billion by the year 2000.[2] In January 1991 the Bush administration boosted funding for its program for biotechnology research to $4 billion.

Great fortunes have already been made by some CEOs —and great fortunes have been lost. Some companies have gone broke, and others have been taken over by larger multinationals. Funded by wealthy corporations, many universities have established biotechnology centers but, ironically, many have lost good teachers and researchers to private industry. A new generation of lawyers has been spawned to deal with the complexities of protecting the "intellectual property rights" of companies developing and marketing new genetically engineered products, and many costly legal battles have already made headlines in industry publications.

Since the first gene was cloned in 1973, the commercialization of biotechnology has taken off at breathtaking speed (see Table 1.1). It has applications in an almost unlimited range of products and services, including medical, agricultural, and industrial. The possibilities

[1]   *The Gene Factory: Inside the Biotechnology Business* (London: Century Publishing, 1985).

[2]   According to an article by Barnaby J. Feder in *The New York Times* (3 Nov. 1991), entitled "Biotech's Biggest Sales So Far Are on Wall Street," recent developments "have fed a bout of biotech fever on Wall Street. The market capitalization of the nearly 200 publicly traded companies has jumped to about $40 billion, almost double the level a year ago. Biotech companies sold more than $2 billion worth of stock in the first six months of 1991." Annual product sales for 1991 exceeded $4 billion.

There has been a parallel explosion of interest in biotechnology in academia. According to the Philadelphia-based Institute for Scientific Information, the number of scientific papers on biotechnology and applied microbiology proliferated from 511 in 1981 to 2,373 in 1990—an increase of 364 percent.

TABLE 1.1.

MAJOR EVENTS IN THE COMMERCIALIZATION OF BIOTECHNOLOGY

1973	First cloning of a gene.
1974	Recombinant DNA (rDNA) experiments first discussed in a public forum (Gordon Conference).
1975	U.S. guidelines for rDNA research outlined (Asilomar Conference). First hybridoma created.
1976	First firm to exploit rDNA technology founded in the United States (Genentech). Genetic Manipulation Advisory Group started in the United Kingdom.
1980	*Diamond v. Chakrabarty*—U.S. Supreme Court rules that microorganisms can be patented. Cohen/Boyer patent issued on the technique for the construction of rDNA. United Kingdom targets biotechnology for research and development (Spinks's report). West Germany targets biotechnology for R&D (Leistungsplan). Initial public offering by Genentech sets Wall Street record for fastest price per share increase ($35 to $89 in 20 minutes).
1981	First monoclonal antibody diagnostic kits approved for use in the United States. First automated gene synthesizer marketed. Japan targets biotechnology (Ministry of International Trade and Technology declares 1981, "The Year of Biotechnology"). Cetus sets Wall Street record for the largest amount of money raised in an initial public offering ($115 million). Over 80 new biotechnology firms formed by the end of the year.
1982	First rDNA animal vaccine (for colibacillosis) approved for use in Europe. First rDNA pharmaceutical product (human insulin) approved for use in the United States and the United Kingdom.
1983	First expression of a plant gene in a plant of a different species. New biotechnology firms raise $500 million in U.S. public markets.
1984	California Assembly passes resolution establishing the creation of a task force on biotechnology. Two years later, a guide clarifying the regulatory procedures for biotechnology is published.

1985	Advanced Genetic Sciences, Inc., receives first experimental use permit issued by EPA for small-scale environmental release of a genetically altered organism (strains *P. syringae* and *P. fluorescens*, from which the gene for ice-nucleation protein had been deleted).
1986	Coordinated Framework for the Regulation of Biotechnology published by Office of Science and Technology Policy. Technology Transfer Act of 1986 provides expanded rights for companies to commercialize government-sponsored research.
1987	U.S. Patent and Trademark Office announces that nonhuman animals are patentable subject matter. October 19th—Dow Jones Industrial Average plunges a record 508 points. Initial public offerings in biotechnology-based companies virtually cease for 2 years.
1988	NIH establishes program to map the human genome. First U.S. patent on an animal—transgenic mouse engineered to contain cancer genes.
1989	Bioremediation gains attention, as microbe-enhanced fertilizers are used to battle *Exxon Valdez* oil spill. Court in Federal Republic of Germany stops construction of a test plant for producing genetically engineered human insulin. Gen-Probe is first U.S. biotechnology company to be purchased by a Japanese company (Chugai Pharmaceuticals).
1990	FDA approves recombinant renin, an enzyme used to produce cheese; first bioengineered food additive to be approved in the United States. Federal Republic of Germany enacts Gene Law to govern use of biotechnology. Hoffman-LaRoche (Basel, Switzerland) announces intent to purchase a majority interest in Genentech. Mycogen becomes first company to begin large-scale testing of genetically engineered biopesticide, following EPA approval. First approval of human gene therapy clinical trial.
1991	Biotechnology companies sell $17.7 billion in new stock, the highest 5-month total in history. Chiron Corp. acquires Cetus Corp. for $660 million in the largest merger yet between two biotechnology companies. EPA approves the first genetically engineered biopesticide for sale in the United States.

Source: Office of Technology Assessment, 1991.

include new vaccines, enzymes, hormones, and diagnostic kits for humans and animals; new-generation pesticides, herbicide-resistant seeds, even plants that produce their own fertilizer and pesticides and are resistant to frost; new-generation food additives and analogs, such as amino acids and vitamins; and more efficient production of commodity chemicals from cornstarch and wood pulp instead of from petroleum. Biotechnology even has applications in bioelectronics and could be used to develop improved biosensors or new conducting devices called biochips, as well as a whole new armory of bioengineered weapons of mass destruction.

According to a report by the U.S. Office of Technology Assessment (OTA), "a well-developed life science base, the availability of financing for high-risk ventures, and an entrepreneurial spirit have led the United States to the forefront in the commercialization of biotechnology."[3] But Japan is a leading competitor, followed by Germany, the United Kingdom, Switzerland, and France. The OTA report goes on to state:

> In Japan, the Federal Republic of Germany, the United Kingdom, Switzerland, and France, biotechnology is being commercialized almost exclusively by established companies. *The Japanese consider biotechnology to be the last major technological revolution of this century, and the commercialization of biotechnology is accelerating over a broad range of industries, many of which have extensive bioprocessing experience* [emphasis mine]. The general chemical and petroleum companies especially are leaning strongly toward biotechnology, and some of them are making rapid advances in R&D through their efforts to make biotechnology a key technology for the future. In Europe, large pharmaceutical and chemical

---

[3]   *Commercial Biotechnology: An International Analysis* (Washington, DC: U.S. Government Printing Office, 1984).

companies, many of which already have significant strength in biologically produced product markets, are the major developers of biotechnology. Their inherent financial, production, and marketing strengths will be important factors as the technology continues to emerge internationally.

As the report emphasizes, it is the large multinational chemical, pharmaceutical, and petroleum companies that are investing in biotechnology. They are buying up not only smaller upstart biotechnology companies but also small and large seed companies. Jack Doyle, in his book *Altered Harvest*, sees this as a clear move toward corporate monopoly of the world's genetic resources.

But monopoly aside, there are significant health, safety, and environmental concerns associated with this new technology such that the established regulatory structure for other industries is wholly inadequate. Not surprisingly, the U.S. biotechnology industry, aided considerably by the President's Council on Competitiveness (chaired by Vice President Quayle), has vigorously opposed effective and appropriate regulation and oversight by responsible government agencies like the Food and Drug Administration (FDA), U.S. Department of Agriculture (USDA), and Environmental Protection Agency (EPA). On this critical issue of health, safety, and environmental regulation, the 1984 OTA report had this to say:

> The analysis of the effect of health, safety, and environmental regulation on competitiveness in biotechnology was made by determining how restrictive a country's laws would be with respect to marketing biotechnology products and whether there were any uncertainties about their application. The analysis focused on the drug laws for humans and animals and, to a lesser extent, on laws governing the production of chemicals and the deliberate release of novel organisms into the

environment. *In all the competitor countries, there is some uncertainty as to the environmental regulation governing the deliberate release into the environment of genetically manipulated organisms* [emphasis mine].

It is especially disturbing that the FDA has taken little initiative in adopting new safety and labeling rules for genetically engineered foods,[4] and that the Council on Competitiveness has made every effort to limit the regulatory scope of the EPA with respect to the environmental release of genetically engineered products and organisms.

Between 1987 and 1991, there were at least 50 government-approved applications by biotechnology companies (for example, Calgene, Amoco, Monsanto, Ciba-Geigy, Upjohn, Pioneer Hybrid, and Campbell Soups) to release plants engineered to carry certain viral and bacterial genes into the environment for field-testing.[5] Transgenic organisms, according to industry watchers, have been field-tested in Argentina, Australia, Belgium, Canada, Chile, Denmark, Finland, France, Germany, Great Britain, Italy, Japan, Mexico, the Netherlands, New Zealand, Spain, and Sweden.

There are so many unknowns involved in releasing such genetically engineered life forms into the environment that the benefits—of bigger and better than ever harvests of tobacco, cotton, tomatoes, potatoes, corn, and soybeans —might not justify the risks and consequences of their release. Several leading ecologists in the United States have urged much tighter, scientifically based regulatory policy over such environmental releases.[6]

---

[4]   See *A Mutable Feast: Assuring Food Safety in the Era of Genetic Engineering*, vol. 2, no. 3 (New York: Environmental Defense Fund, 1991).

[5]   *The Gene Exchange*, vol. 2, no. 3 (Washington, DC: National Wildlife Federation, 1991), pp. 11–14.

[6]   See L. Roberts, "Ecologists Wary about Releases," *Science*, 3 March 1991, p. 1141.

The biotechnology industry justifies its laissez-faire attitude toward the environmental and social consequences of new products and life forms (which is shared by many politicians and bureaucrats) on the grounds of economics and world-market competition. But the serious adverse reactions of many people after consuming an FDA-approved genetically engineered food supplement (the amino acid L-tryptophan), which the regulators thought to be safe, should be warning enough. Some 27 people died and 1,500 were seriously injured.[7]

The Money Game

The life science departments of most universities have joined the scramble for research funds in biotechnology, so it is understandable (but not excusable) that some researchers in academia are eager to jump on the bandwagon and even engage in field tests before receiving official approval.

Federal and state support of biotechnology research has mushroomed over the past decade. In 1987, according to the OTA,[8] state support of universities totaled $147 million, and federal agencies spent some $2.7 billion (of which $119 million went to the Department of Defense). The OTA identified 403 U.S. companies involved in biotechnology and estimated that they were spending some $1.5 to $2 billion annually in research and development (R&D). The primary areas of R&D focus are summarized in Table 1.2. By 1990, U.S. taxpayers were providing

---

[7]   See E. A. Belongia et al., "An Investigation of the Cause of the Eosinophilia-myaglia Syndrome Associated with Tryptophan Use." *The New England Journal of Medicine* 323 (1990): 357–365.

[8]   OTA Report Brief, Washington, DC, July 1988.

TABLE 1.2.

AREAS OF PRIMARY R&D FOCUS BY BIOTECHNOLOGY COMPANIES

Research Area	Dedicated Biotech Companies	Large Established Companies
Human therapeutics	63 (21%)	14 (26%)
Diagnostics	52 (18%)	6 (11%)
Chemicals	20 (7%)	11 (21%)
Plant agriculture	24 (8%)	7 (13%)
Animal agriculture	19 (6%)	4 (8%)
Reagents	34 (12%)	2 (4%)
Waste disposal/treatment	3 (1%)	1 (2%)
Equipment	12 (4%)	1 (2%)
Cell culture	5 (2%)	1 (2%)
Diversified	13 (4%)	6 (11%)
Other	31 (18%)	0 (0%)

Source: Office of Technology Assessment, 1988.

various federal agencies with almost $3.5 billion for bio-technology R&D (see Table 1.3). Public funds are deemed necessary if the United States is to maintain a competitive edge in the world market, Japan being its primary rival (see Table 1.4). According to the Worldwatch Institute,[9] most of the money invested in biotech research comes not from public or philanthropic organizations but from multinational corporations. In 1987, U.S. industry spent $300 million and the U.S. Department of Agriculture $85 million, and agribiotech record.

Considerable public funds are being used by government research institutions and by state universities to research and develop new biotechnology products, techniques, and services, which will then be picked up by private industry

[9]   *State of the World* (New York: W. W. Norton & Co., 1990).

TABLE 1.3.

U.S FEDERAL FUNDING FOR BIOTECHNOLOGY, FISCAL YEAR 1990 (millions of dollars)

Agency	Amount
National Institutes of Health	$2,900.0
National Science Foundation	167.9
Department of Agriculture	116.0
Department of Defense	98.0
Department of Energy	82.2
Agency for International Development	28.7
Food and Drug Administration	19.4
Environmental Protection Agency	8.3
Veterans Administration	7.5
National Institute of Standards and Technology	4.8
National Aeronautics and Space Administration	4.5
National Oceanic and Atmospheric Administration	2.0
Total	$3,439.3

Source: Office of Technology Assessment, 1991.

for commercial development and marketing. This becomes a point of concern when public-funded R&D in biotechnology advances the private interests of corporations rather than serves the public interest. This is especially true for expenditures associated with developments in agricultural biotechnology. Land-grant schools could do a lot to help improve the economic well-being of rural communities and protect the family farm tradition from total extinction. Instead, most developments in agriculture, facilitated by universities, have hastened the demise of local farming communities. The direction being taken in R&D in agri-biotechnology does not bode well for the good farmer (see

TABLE 1.4.

BIOTECHNOLOGY COMPANIES AND SECTOR OF INTEREST

Source	USA	Japan	France	UK	Federal Republic of Germany
Agriculture	73	12	5	15	2
Chemical	37	31	1	4	4
Diagnostics	141	15	3	10	6
Food	18	17	2	12	1
Pharmaceuticals	65	28	2	9	4
Veterinary	54	2	3	6	0
Total	388	105	16	56	17

Source: Organization for Economic Cooperation and Development, 1988.

Chapter 4 for more details). According to the OTA, the USDA received $84 million in federal funds for biotechnology research in 1987, and this massive infusion of public funds continues. Yet in 1991, the USDA received a paltry $6.7 million for R&D into alternative, sustainable agriculture.

Risks and Benefits

Humankind now has power over the genes of life. Any new technology brings with it costs and risks as well as benefits.

The benefits of genetic engineering will be assured only if the potential risks and costs are fully acknowledged and international regulations are instigated to ensure optimal benefit for all rather than maximal benefit for a few.

The risks to animals are considerable—from suffering as a consequence of genetic manipulation, to habitat competition and loss following the deliberate or accidental release of genetically engineered life forms. The non-medical risks to humans are environmental, social, and ethical, and are spawned by a rising technocracy of *genetic imperialism*.

For example, this new technology could be used to screen people and deny some the right and access to equal and fair employment or to health and life insurance. It is already a threat to the agricultural economies of some third-world countries. One example is the synthetic production of aromatic oils and essences like vanilla in the United States, which is in direct competition with African and other third-world countries whose economies are dependent on the export of the natural forms of these products.

Properly applied, genetic engineering could be directed to help improve the resilience and nutritive value of crops, to develop vaccines against tropical diseases, and to enhance the quality of life and security of half the world's population. Four billion livestock along with 5.3 billion people are a major threat to the Earth's ecology. If livestock were healthier and more productive, fewer would be kept. Improved health and life expectancy of people would help population control in the third world. Studies have shown that couples have many children because they expect high infant mortalities, and they need children to help farm the land and care for them in their old age. Improved quality of life and economic security would help reduce this need for large families, provided there was appropriate education, family planning, and access to safe and effective birth-control methods. New genetically engineered vaccines to induce temporary infertility are especially promising.

The Worldwatch Institute has noted that:

U.S. biotechnology companies are fond of describing miracle crops that will feed hungry millions, and often use the promise of such immense benefits to attract investors and justify minimal regulation. But the application of biotechnology in the rich nations may not yield the rather prosaic miracles needed by small farmers in developing countries. Corporations will use biotechnology to develop crops and agricultural products they expect to be profitable and, if possible, which they can patent.[10]

Current trends in some agricultural applications of new technology are cause for concern. These include the development of crop varieties genetically engineered to be resistant to pesticides and herbicides and genetically engineered vaccines and other products, such as bovine growth hormone, to increase the productivity of intensively raised farm animals.

From humane, economic, and environmental perspectives, a reduction in the numbers of livestock and an end to the intensive ways in which they are raised in developed countries are indicated. Genetic engineering is being misapplied when it is directed toward the continuance of chemically addicted agriculture and the perpetuation of unhealthy, drug-dependent, environmentally unsound, and often cruel livestock and poultry husbandry practices.

Farm animals that carry certain human genes are now being turned into biochemical machines to produce biological materials in their milk and blood for the pharmaceutical industry. One veterinary biotechnologist called this new development "molecular farming" and, in testimony before the House Judiciary Committee, presented it as the salvation of U.S. agriculture and the family farm. Legislators, no less well-intentioned than the dedicated

[10] *State of the World* (New York: W. W. Norton & Co., 1990).

gene engineers, believe that industrial productivity and economic growth are vital to national security and world-market supremacy. They will help the biotechnology industry grow. But their children's children will inherit what they are unwittingly helping to create today: the silent world of tomorrow, save for the clamor of industrious man and his own inventions. A denatured world and analog foods should not be their legacy.

Before meat is ever recreated in analog form in the biotech industrial laboratory, genetically engineered pigs, cows, and chickens will be fed with genetically engineered supercorn and soybeans. Farm animals and crops alike will be made resistant to disease and made ever more productive, but to what end? Will the world be a better place and people better fed with these fruits of biotechnology? As I document in subsequent chapters, the overall direction being taken by the biotechnology industry is contrary to the long-term interests of the global community and to the integrity and future of creation. The primary beneficiaries will be the biotechnology industry and a new elite that has a monopoly over the genetic resources of the world. And in the process of gaining control over life for purely commercial purposes, they will cause more harm than good.

This direction can be changed to result in more good than harm, but not without the involvement of an informed and concerned public with a diversity of views, values, and concerns. It is not that the scientists involved in genetic engineering or their funding agencies, investors, lobbyists, and corporate boards have some hidden agenda or harmful intent. Rather, they are so fixated on their own narrow goals that they fail to see the broader ramifications and potentially harmful consequences of their endeavors.

As Prof. Ian G. Barbour observes, "The control and direction of technology involves ethical values such as justice, freedom, and environmental stewardship. Respect for persons and for nature is not a scientific conclusion;

wisdom in applying knowledge torward humane goals is
not a product of the laboratory."[11]

Our newly acquired power over the gene is a consequence of the *collective* intelligence of our species acquired
over millennia. It is a power and knowledge that should
not be monopolized by a minority (which I euphemistically
refer to as the biotechnocracy). Rather, it should be shared
by a global *biodemocracy* that applies its new-found power
and knowledge to serve the greater good; to restore and
heal the Earth; and to dress and keep the natural world as
best we can.

[11]   Ian Barbour. *Religion in an Age of Science*. The Gifford Lectures,
       vol. 1 (San Francisco: Harper, 1990).

GENETIC ENGINEERING AND SCIENTIFIC IMPERIALISM

Thou shalt not let thy cattle
gender with a diverse kind;
Thou shalt not sow thy field
with mingled seed.
—LEVITICUS 19:19

What do most scientists and doctors think about genetic engineering now that they can put the genes of one species into another—so-called transgenic manipulation? They have already succeeded in inserting human genes into mice, cattle, pigs, salmon, and sheep. This new technology, which gives us the power to control and redirect the entire evolutionary process, raises fundamental ethical and moral questions. One central question concerns the ethics and morality of disrupting the inherent nature of animals—their *telos* (a Greek word meaning "end" or "aim")—for purely human ends. Transgenic manipulation between species can produce far more profound changes in a species than have ever been possible through selective breeding within species. Is this a violation of the dignity of nonhuman life? What of the long-term consequences to the animal kingdom, to the environment, and to nature's creative process?

Views of the Scientific Community

I have never felt more alienated from my own kind when, in the spring of 1985, I confronted the National Institutes of Health Genetic Engineering Committee in Washington, D.C. Jeremy Rifkin of the Foundation on Economic Trends and I challenged the National Institutes of Health to temporarily suspend government-funded transgenic animal research until the ethics and consequences of developing new industries out of this biotechnology had been fully explored and publicly aired. We were met with united opposition. In the committee's large conference room— with scientists gathered around a thirty-foot-long oval table and the press and observers seated around them— I experienced a sense of vertigo and unreality as the chairman read statements from scientists supporting transgenic research. These statements came in rebuttal to the ethical question that Rifkin and I had raised about the rightness of interfering so profoundly with the inherent nature of animals through such manipulation. One statement implied that this was a perfectly natural development in human evolution: to play God. Another insisted that animals have no inherent nature (i.e., no intrinsic worth) because their *telos*—or final end—is death.

Thus it was reasoned that there was nothing wrong with directing that final purpose to satisfy purely human ends.

That none of the academically esteemed scientists and bioethicists on the committee questioned these assumptions and this wholly instrumental and self-serving attitude toward life was the most shocking experience I have had in my career as a scientist and spokesperson for animal welfare and rights.

Here is a selection of statements by various scientists and physicians that were handed out at this unforgettable meeting:

> The idea that a species has a "telos" is contrary to any evidence provided by biology and belongs rather in the realm of mysticism. That mysticism is a poor basis for sound public policy is amply confirmed by history. (PROFESSOR M. J. OSBORN, Chairman, Department of Microbiology, School of Medicine, University of Connecticut)

> History, from Galileo through Lysenko, teaches us that mysticism can never yield rational and wise public policy in scientific matters. . . . The notion that a species has a telos (a purpose) contravenes everything we know about biology. Species can have, and many in the past have had, a telos (an end), namely, extinction. That is the only telos known to exist. (DR. MAXINE SINGER, National Institutes of Health)

These scientists dismiss philosophy and ethics as mysticism, failing to grasp the full and original meaning of *telos*.[1]

---

[1] Aristotle recognized that form and matter coexist and contended that the "cause" of any changes in these can be subdivided into four parts: (1) formal cause, which is the essence or pattern in matter that persists as the matter undergoes change; (2) material cause, which is the substratum, the matter in which the essence or form is contained; (3) efficient cause, which refers to the proximate agents

But the debate about *telos* is a matter of semantics. The real issue is whether living things have inherent natural qualities that we tamper with at our peril. I believe that they do. If this is mysticism, so be it.

Another biomedical scientist who wrote to the National Institutes of Health in support of transgenic research presented an evolutionist's rationale for transgenic manipulation—a kind of biological determinism:

> We humans are participating in the process of evolution *per se*. By that I mean that our ability, acquired through evolution, to manipulate genomes by selective breeding and more recently by recombinant DNA technology is an integral component of evolution itself and is not, as has been claimed in the past, "tinkering with evolution." Instead, *it is* evolution. Because of our evolved level of consciousness, we as human beings must, however, utilize appropriately our consciousness and ability to anticipate as we effect our role in the evolution of our own and other species on this planet. Thus, I do not subscribe to the premise that we as human beings must through regulatory agencies exercise control over evolution by forbidding specified acts of nature, we being mere agents of nature. (DAVID W. MARTIN, JR., M.D., School of Medicine, University of California, San Francisco)

Yet other supporters of transgenic research argued that it is ethically acceptable because we have been crossbreeding and selectively breeding animals for centuries (ignoring

---

of change; and (4) final cause, or the purpose for which the change was initiated. *Telos*, defined in *Webster's Third International Dictionary* as "an ultimate end or object," includes all four of these subdivisions, a point often overlooked by those who regard *telos* as referring only to final cause. Since *telos* also relates to the qualities of an object (formal cause and material cause), it is correct, as philosopher Bernard Rollin has insisted, to infer that the term *telos* refers to an inherent nature or beingness and not simply to the ultimate end or final purpose of an object or living entity.

the data showing that such manipulations have increased animal disease and suffering through genetically related and "domestogenic" diseases). For instance, Professor E. Brad Tompson, Chairman of the Department of Human Biological Chemistry and Genetics at the University of Texas Medical Branch at Galveston, wrote:

> The truth is that man has been experimenting with crossing animal species since time immemorial. The technology available to do it now simply differs from that available formerly. It is, in my opinion, dangerous and wrong for a prohibition of the sort suggested to be put into place as part of the framework in which American research is conducted. It would undoubtedly deter important and potentially useful experiments from being done, experiments which would have potential for improving the lot of many species including but not limited to mankind.

Some scientists seem to believe that there is no real difference between the traditional techniques of selectively breeding domestic animals for certain useful traits and the new engineering technique of inserting "useful" genes of other species. The latter is seen simply as an extension of the former. For example, Dr. Cornelius Van Dop of the Johns Hopkins Hospital, Baltimore, wrote:

> The selective breeding of animals directed to amplifying or eliminating certain traits has been a human activity since the first mammal was domesticated during prehistoric times. This selection for specific traits (mutated genes) has irreversibly modified the gene pools of innumerable species for man's economic gain and whim. . . . Current bioengineering technology stands at the threshold of being able to selectively modify one gene at a time and thereby reduce dependence on selective breeding for altering

certain traits. The selective introduction of foreign genes into germ lines is thus a logical extension of animal husbandry.

There is a world of difference between genetic engineering and selective breeding. Transgenic manipulation entails crossing the natural biological boundaries *between* animal species. This has *not* been done before. (The closest analogy is the crossbreeding of closely related subspecies such as horses and donkeys or wolves and dogs.) Single genes can have profound consequences, and increasing the utility of pigs with cattle growth genes, or cattle with elephant growth genes, could so disrupt the animals' *telos* (intrinsic nature) biophysically, metabolically, and developmentally as to create a host of health and welfare problems that would require further technological "fixes" with chemicals, drugs, and additional genetic manipulation.

Jeremy Rifkin observes that:

> Already researchers in the field of molecular biology are arguing that there is nothing particularly sacred about the concept of a species. As they see it, the important unit of life is no longer the organism, but rather the gene. They increasingly view life from the vantage point of the chemical composition at the genetic level. From this reductionist perspective, life is merely the aggregate representation of the chemicals that give rise to it and therefore they see no ethical problem whatsoever in transferring one, five or a hundred genes from one species into the hereditary blueprint of another species. For they truly believe that they are only transferring chemicals coded in the genes and not anything unique to a specific animal. By this kind of reasoning, all of life becomes desacralized. All of life becomes reduced to a chemical level and becomes available for manipulation.[2]

[2]   J. Rifkin, *Algeny: A New Word—A New World* (New York: Viking, 1983), p. 47.

Dr. David Baltimore, former director of the Whitehead Institute for Biomedical Research, Cambridge, Massachusetts, expressed the technocrat's instrumental ideology of separating science from ethics and morality (because the latter are subjective), and of leaving ethical decisions, such as transgenic manipulation, to rationalism:

> I oppose "prohibitions" on the grounds that they provide an apparent simplicity that often leads to difficulty. I also oppose writing into regulations statements about "morally and ethically un-acceptable" practices because those are subjective grounds and therefore provide no basis for discussion. There are good scientific grounds for not putting any new genes into the human gene line today and I believe that we should rest our behavior on such rational assessments not on the shifting and personal grounds of morality.

It is ironic that most scientists and select committees believe that although there is no ethical issue involved in switching genes between animal species and putting human genes into animals, it is wrong for animal genes to be put into human gene lines. Clearly, only human life is sacred.

Other letters from biomedical scientists emphasized that transgenic research is essential for understanding how genes work and can be controlled so that genetic and developmental defects in children might be corrected. But they miss the point that medical genetic engineering is interventive rather than preventive, and since genetic and developmental defects are in part environmentally correlated (e.g., with teratogenic and mutagenic agrichemicals, food additives, industrial chemicals, and pollutants), genetic engineering will be no panacea if environmental factors continue to be ignored for political and economic reasons. Rather, it will be yet another lucrative medico-technical fix upon which an ever-sickening society will become dependent.

The potential benefits of genetic engineering in the eyes of biotechnologists seem to have no limits, especially when attention is focused on farm animals. In the worldwide laboratory of genetic engineering, farm animals are the principal guinea pigs after the mouse, which has a veritable zoo of transgenic varieties. Professor C. Eugene Allen, Dean of the College of Agriculture at the University of Minnesota, stated:

> A fecundity gene has been identified in a flock of Merino sheep in Australia which, if successfully transferred to cattle, could increase the number of twin compared to single births. Such a breakthrough would have a major impact in reducing the cost of producing beef in the U.S. and many countries where feed for cattle is not a limiting factor.

Many scientists have endorsed transgenic research without question, ignoring the environmental impact on and displacement of indigenous wildlife species that would inevitably follow an increase in the worldwide population of farm animals. They adhere to the myth that genetic engineering will make these animals even more efficient and productive, forgetting that you can't get something for nothing. Professor Neal L. First, Department of Meat and Animal Science, University of Wisconsin, Madison, stated:

> Our research presently concerns the introduction of genes of other species into the germ line of food-producing species and the multiplication of the resulting zenogenous embryos. We believe this research will, for example, through the introduction and exogenous regulation of a foreign growth hormone gene, provide genetic stocks which require 25% to 30% less food to produce a

pound of meat and are capable of at least 15% more milk production. Engineering the genes of rumen microorganisms to digest cellulose and lignin will mean that cows, sheep and water buffalo in world land areas of human starvation can convert branches of trees, brush, weeds and fibrous plant residues to needed human food.

The introduction of exogenous genes from species resistant to diseases is expected to allow the use of food-efficient or high-producing livestock in areas of the world where they might not normally survive.

This will lead to further displacement of wildlife and extinction of species. Thomas E. Wagner, Professor and Chairman of the Department of Molecular and Cellular Biology, Ohio University at Athens, wrote:

Growth hormone genes (of human origin) are being transferred into the germ lines of swine and other livestock animals in our laboratory and in the laboratories of others because the resulting animals *will probably be dramatically more feed efficient* and will provide agriculture with some of the economic benefits of the biotechnology revolution. The Ohio Farm Bureau is a partner in this research undertaking in Ohio because this technology may serve to aid in improving the depressed farm economy.

Veterinarians Richard D. Palmiter and Ralph L. Brinster from the School of Veterinary Medicine, University of Pennsylvania, also envisage a new use of genetically engineered farm animals as biochemical machines from which various biologics could be "harvested." They stated:

Another area of research involving transfer of genes into mammals that could lead to beneficial results involves the genetic manipulation of farm animals. There are two major categories of ideas that are being considered. One involves introducing genes into farm animals that would confer upon them desirable traits, e.g. increased growth, milk production or resistance to disease. The other involves introducing genes that code for proteins of medical value. The idea is that one could harvest the protein of interest from the blood or milk of these animals just as one would do from a bacterial or yeast culture. The idea warrants consideration because there are a number of valuable gene products (e.g. certain blood clotting factors) that are functional only after special modifications which only occur in mammalian cells.

Alternative Views

As Canadian philosopher Alan R. Dregson observes: "The technocrat does not question the human right to control nature."[3] Problems that arise from invoking that "right" are seen as being caused by a lack of scientific knowledge rather than a lack of ethics. And finding the solutions justifies more research and technological "fixes." Ethical sensibility is supplanted by hubris, economic determinism, and the erroneous belief that anything that profits humanity is good, regardless of the negative impact on nature and the animal kingdom. The destruction of nature, extinction of species, and the suffering of animals, which are inevitable consequences of their industrialized exploitation by the chemical, biomedical, and agribusiness industries, are justified on the grounds of unavoidable necessity and

[3] *Zygon* 19(1980): 259–275. See also A. R. Dregson, "Shifting Paradigms from Technocratic to the Person-Planetary." *Environmental Ethics* 2(1980): 221–240; and D. Ehrenfeld, *The Arrogance of Humanism* (New York: Oxford University Press, 1978).

the price of progress. Correctives are instigated only if they are economically justifiable.

Even the public's health and the rights of consumers to wholesome food, clean air, and clean water are sacrificed. In order to preserve the status quo of economic determinism, new medical and legislative fixes are found to correct these residual problems rather than change values and practices to correct the underlying problem. For example, applying poisonous pesticides and other agrichemicals is profitable, as are the medical procedures necessary to treat cancer, birth defects, sterility, and other diseases linked with these agripoisons. Preventive medicine is as unprofitable to the medical industry as organic farming is to agribusiness—but certainly not to the consuming populace. It is disturbing to contemplate the irony of applying genetic engineering technology to correct a host of human health problems that are due in part to the misapplication of chemical and medical technologies in agriculture and farm animal production. Genetic engineering in human medicine is not really progressive because its primary focus is not preventive; it is essentially yet another profitable, alternative, interventive technological fix, the primary beneficiaries of which will be the investors, manufacturers, hospitals, and medical administrators—not the public.

Although the blossoming medical and agricultural genetic engineering industries may help us and our plants and animals adapt to increasingly pathogenic environmental conditions, it would surely be more prudent to clean up the environment. This is well recognized as a fundamental principle of preventive "holistic" ecological medicine and of sustainable agriculture, but in our technocracy, for economic and political reasons, it is ignored by organized medicine and relegated at great public expense to ineffectual government regulatory agencies such as the FDA, EPA, and USDA. The difference between democracy and technocracy is that the latter sacrifices public interest for private interest under the guise of progress and the greater good of society. As such, it is a species of imperialism that claims to be altruistic but is actually self-serving and self-perpetuating.

No structure exists outside of Congress to permit public involvement in the policy- and decision-making processes whereby a new technology, whether it be nuclear power or genetic engineering, can be safely developed or prohibited. The public has been told to have faith in science—the religion of materialism—and that its technocrats know best. What we need are not new technological fixes and medical miracles, but rather a fundamental change in worldview. Without a resacralization of nature and compassion and humility toward all living things, we will lack the ethical and moral constraints to use our power over life, over the atom, and over the gene nondestructively, in harmony with the laws of nature and the creative process of which we are an integral part, not lord and master.

CHAPTER 3

REGULATING BIOTECHNOLOGY

I'm drawing up the Whole Risk catalog. Under D, I have dogs, doctors, dioxin. Where do I put DNA? Very low.
—JAMES D. WATSON, THE DNA STORY: A DOCUMENTARY HISTORY OF GENE CLONING

*F*rancis Bacon cautioned that if we would control nature, we must first obey her. Today we understand obedience as respect for nature's "laws." The most important aspect of this respect is being responsible for the environmental and ecological consequences of our actions. This has now become written into our law, which, for better or worse, is administered by the Environmental Protection Agency (EPA.)

This agency recently expressed confidence that the safety and environmental impact of genetically engineered organisms can be adequately determined and regulated. Such confidence is undermined by the EPA's historically documented inability (along with the U.S. Department of Agriculture) to protect the environment and the public's health from the wholesale misapplication of petrochemical-based agripoisons (pesticides, herbicides, and fungicides).[1] Although these old agrichemicals now contaminate our food, water, and body tissues, at least they did not have the capacity to multiply; the new bacterial pesticides do.

Clearly genetic engineering is as much a Pandora's box as it is a cornucopia of wonderful possibilities. Like any other product of human ingenuity, it has great potential risks as well as benefits to society. And for those who invest in this new industry—since genetically engineered plants and microorganisms can be patented—fortunes can be made. But without congressional and state oversight and international coordination to minimize risks to the environment and to the very fabric of life itself, we could be on the threshold not of some biological utopia, but of our own nemesis.

The Watchdogs

The blossoming biotechnology industry is now regulated by already existing governmental regulatory agencies that have been jockeying for jurisdiction over various sections of the industry (see Table 3.1). But how effective can these agencies be?

---

[1]   For example, see Lewis Regenstein, *America the Poisoned* (Washington, DC: Acropolis Press, 1982).

TABLE 3.1.

FEDERAL AGENCIES RESPONSIBLE FOR THE APPROVAL OF BIOTECHNOLOGY PRODUCTS

Biotechnology Product	Responsible Agency
Pesticide microorganisms released in the environment	EPA APHIS
Other uses of microorganisms: Inter-generic combination	EPA APHIS
Foods/food additives	FDA FSIS
Human drugs, medical devices, and biologics	FDA
Animal drugs	FDA
Animal biologics	APHIS
Other contained uses	APHIS FSIS FDA

APHIS: Animal and Plant Health Inspection Service
EPA: Environmental Protection Agency
FDA: Food and Drug Administration
FSIS: Food Safety and Inspection Service

The present governmental regulatory framework for the U.S. biotechnology industry is based on the Federal Policy on Biotechnology. Dr. James W. Glosser, chief administrator of the U.S. Department of Agriculture's (USDA) Animal and Plant Health Inspection Service (APHIS), summarized it as follows:

The Federal Policy on Biotechnology was established December 31, 1984, and published in final form on June 26, 1986, by the U.S. Office of Science and Technology Policy (OSTP) as the "Coordinated Framework for Regulation of Biotechnology." The OSTP concluded that products of recombinant DNA technology will not differ fundamentally from unmodified organisms or from conventional products. Therefore, the existing laws and programs are adequate for regulating organisms and products developed by this process.[2]

Many would not agree with this conclusion, which is the cornerstone of the U.S. government's oversight and regulation of the biotechnology industry. Dr. Glosser went on to state:

> The Coordinated Framework included an index of laws applicable to biotechnological products in the various stages of research, development, marketing, shipment, use, and disposal. The framework also included policy statements from the three federal agencies that share the major responsibilities for regulating products of recombinant DNA technology—the Food and Drug Administration, the Environmental Protection Agency, and the USDA. . . .
>
> A key element in our structure was creation of a special staff to coordinate biotechnology regulatory activities for the Department. The staff serves as a liaison with other government agencies, industry, and the general public on USDA's regulation of biotechnology. The staff actively participates in the United States effort to promote international consistency on biotechnology regulations. Such international consistency should prevent unnecessary trade barriers and ease the transfer of American products into international markets.

It is noteworthy that the emphasis here is on promoting international regulatory consistency not to minimize

---

[2]   In testimony before the House Subcommittee on Department Operations, Research and Foreign Agriculture on Oversight of APHIS Programs, 17 July, 1988.

adverse environmental consequences, but rather to facilitate the global expansion of U.S.-based biotechnologies.

After a year's study of biotechnology regulation at the USDA, the Government Accounting Office reached three main conclusions in its report released on 3 April 1986:

1. The main biotechnology committee at the USDA —the Agriculture Recombinant DNA Research Committee—has almost no authority, meets infrequently, has no budget, and its meeting records show confusion about what the committee should do.

2. The department has no clear policy about who should review biotechnology proposals and what rules should be applied. The report stated, "Different agencies in USDA have been jockeying for regulatory control, and USDA officials have expressed uncertainty as to which agency is responsible for different activities."

3. The USDA has done little to communicate to Congress and the public the benefits and risks of biotechnology.

The mandate of this chronically understaffed and underfunded federal agency (the USDA) to protect animal and plant health clearly omits reference to environmental protection, which is the domain of the EPA. Some would argue that the EPA should regulate all environmental releases of biotechnology products, since the USDA was so ineffectual in regulating the pesticide industry that the responsibility was transferred to the EPA.

How can the USDA be expected to effectively regulate the release of engineered plants, microorganisms, and animals (including insects and other plant pests) as well as animal drugs and live-virus vaccines developed through genetic engineering, considering its past failure with pesticides? But it is doubtful that any federal agency can effectively regulate this new industry.

The Food and Drug Administration (FDA) regulates new drugs and vaccines developed from biotechnology. But the existing regulations and test protocols for all new pharmaceuticals are inadequate and inappropriate according to a recent Government Accounting Office report. For example, growth hormone manufactured from genetically engineered bacteria passed all the routine assays and toxiology tests but had some unanticipated and significant clinical side effects, which were subsequently found to have been caused by unidentified contaminants. The public has not forgotten the thalidomide tragedy (when animal tests run before its approval for human use proved, retrospectively, to be inappropriate and invalid) nor the recent tragedy caused by genetically engineered L-tryptophan that was thought to be safe.[3]

The FDA is also in charge of monitoring our food for contamination with agrichemicals, drugs, and bacteria, along with the USDA's meat inspectorate. Data indicating high levels of agrichemical residues in imported and domestic foods cast doubt on the FDA's ability to regulate the agribusiness food industry effectively if and when biotechnology, as well as food irradiation, become integral components.

The regulatory circus and bureaucratic gridlock of finding effective ways of governing the safe and appropriate applications of biotechnology in the United States have been succinctly summarized by the Office of Technology Assessment (OTA) as follows:

---

[3]   *Food Safety and Quality: FDA Needs Stronger Controls Over the Approval Process for New Animal Drugs* (Washington, DC: General Accounting Office, 1992).

The Biotechnology Science Coordinating Committee (BSCC) was founded by the Office of Science and Technology Policy (OSTP) in 1985 to:

> ...serve as a coordinating forum for addressing scientific problems, sharing information, and developing consensus; promote consistency in the development of Federal agencies' review procedures and assessments; facilitate continuing cooperation among Federal agencies on emerging scientific issues; and identify gaps in scientific knowledge.[4]

The committee consisted of the Commissioner of the FDA, the NIH Director, the Assistant Secretary of Agriculture for Marketing and Inspection Services, the Assistant Secretary of Agriculture for Science and Education, the Assistant Administrator of the EPA for Pesticides and Toxic Substances, the Assistant Administrator of the EPA for Research and Development, and the Assistant Director for Biological, Behavioral, and Social Sciences of the National Science Foundation.

Rather than being a forum for discussion, however, BSCC became the center of interagency disagreements about regulatory policy. Internal dissension reached a climax in 1988, when EPA sent its proposed rule for regulation of genetically modified micro-organisms under TSCA to the Office of Management and Budget (OMB) for review before publication in the Federal Register. The chairman of BSCC wrote to OMB requesting that OMB withhold clearance until BSCC could consider the proposed rule. A series of interagency meetings and memoranda resulted in deadlock. The chairman informed OMB, and OMB refused to approve EPA's draft rule. In response, the EPA representative to BSCC stopped attending meetings and placed the draft rule and interagency

---

[4]   *Biotechnology in a Global Economy* (Washington, DC: U.S. Government Printing Office, 1991).

memoranda in a public docket. As of mid-1991, no proposed rules for EPA regulation of micro-organisms under TSCA and FIFRA had been published.

One major area of disagreement was the pre-cise definition of organisms that would be subject to EPA regulations. In 1989, various approaches to this problem were discussed by BSCC and by the agencies' scientific advisory committees. Not surprisingly, BSCC failed to reach a consensus. The issue was turned over to a higher level com-mittee, the Biotechnology Working Group of the President's Council on Competitiveness, chaired by Vice President Quayle. The OSTP's proposed principles for the scope of oversight for the planned introduction of organisms were published in July 1990.

In late 1990, BSCC was replaced by the Bio-technology Research Subcommittee (BRS) of the Committee on Health and Life Sciences, a standing interagency committee of the Federal Coordinating Council on Science, Engineering, and Technology (FCCSET). The FCCSET, like OSTP, is headed by the President's science advisor. The BRS's charge is said to be similar to that of BSCC. Its mem-bership is broader and includes representatives from the Department of Energy (DOE), NIH, FDA, the State Department and its Agency for International Development (AID), EPA, USDA, NSF, the National Aeronautics and Space Admin-istration (NASA), the Department of Commerce (DOC), the Department of Defense (DoD), the Department of the Interior, OMB, and OSTP.

This OTA report, in addition to revealing the regulatory disarray of biotechnology, emphasizes that the EPA has been regulating the release of genetically engineered microorganisms intended for use as pesticides under the Federal Insecticide, Fungicide and Rodenticide Act (FIFRA) and for other purposes under the Toxic

Substances Control Act (TSCA). Both these acts were intended to regulate chemical pesticides, not living, genetically engineered life forms. Many analysts consider them wholly inadequate for this purpose[5], and I wholly concur. According to the OTA, the EPA now works with APHIS in reviewing industry and university applications under FIFRA to release microorganism pesticides. As of March 1991, EPA had approved ten applications for small-scale testing of genetically engineered microbial pesticides under FIFRA. In addition, two applications had been withdrawn, and another review had been suspended.

The Recombinant DNA Advisory Committee of the National Institutes of Health has established guidelines for research. But some of the committee members believe that the existing guidelines should be abolished because there have been no accidents involving genetically engineered bacteria and the public health in the fifteen years since the committee's inception. What is more disturbing is that this same committee ignored its own guidelines by permitting University of California agriculturalists to release genetically engineered bacteria into the environment.

Cornell University professor David Pimental is one outspoken ecologist who has expressed concern over the continuing lack of coordination between various government regulatory agencies and the lack of sound protocols for the release of new life forms into the environment. He writes:

---

[5]   For example, see S. Krimsky et al., "Controlling Risk in Biotech." *Technology Review* 92(5): 62–70 (1989); G. Marchant, "Modified Rules for Modified Bugs: Balancing Safety and Efficiency in the Regulation of Deliberate Release of Genetically Engineered Organisms." *Harvard Journal of Law and Technology* 1 (Spring 1988): 163–208.

Setting up regulatory procedures is a much-needed first step in controlling such [agricultural] activities. In June 1986, the federal government approved rules and guidelines for regulating the biotechnology industry. Responsibility for weighing the safety of new products was divided among five federal agencies:

- The USDA is responsible for engineered organisms used with crop plants and animals.
- The FDA is responsible for genetically engineered organisms in foods and drugs.
- The National Institutes of Health are responsible for engineered organisms that could affect public health.
- The Occupational Safety and Health Administration (OSHA) is responsible for engineered organisms that may affect the workplace.
- The EPA is responsible for engineered organisms released into the environment for pest and pollution control and related activities.

However, many scientists believe that these divisions of authority are cumbersome and inadequate. Agencies such as the USDA both promote and regulate the new technology. That combination did not work when the USDA was responsible for the use of pesticides, so control of those substances was transferred to the EPA in 1970. In my opinion, biotechnology should be regulated primarily by the EPA and OSHA—agencies that are set up to regulate rather than promote industries. The other agencies could contribute representatives to the EPA and OSHA committees that would review the release of engineered organisms.

These agencies should ensure that sound ecological protocols are followed before any genetically engineered organism is released into the environment.[6]

On 21 May, 1986, the White House Domestic Policy Council approved a long-awaited document that established rules for regulating the biotechnology industry. This document was formulated by the White House's now defunct Biological Sciences Coordinating Committee (BSCC) chaired by Dr. David T. Kingsbury. Philip J. Hilts reported the following in *The Washington Post* (22 May, 1986):

> Kingsbury said the new policy establishes a framework under which all federal agencies would regulate biotechnology products. Some products would be exempted and others assigned for regulation by specific agencies, preventing "agency shopping" by manufacturers.
> The document states, in effect, that each product should be considered separately and not be subject to regulation solely because it was made by genetic engineering techniques, he said.
> This position was sought by industry and opposed by some critics and environmental groups.
> One of the document's most controversial provisions exempts a large group of genetically engineered products from additional regulation.
> For example, experiments that simply cut genes from an organism, a process called "deletion," will be exempt from regulation. Three of the first four products to be reviewed at the EPA and the USDA are "deletion" cases that probably would not undergo regulatory review unless found to fall into some other regulated category, such as organisms that may cause disease.

[6] "Down on the Farm: Genetic Engineering Meets Ecology," *Technology Review*, January 1987: 24–30.

In the fall of 1987 a provisional report by the OTA, highly critical of the government's ability to adequately regulate the biotechnology industry, was released for public comment before the preparation of a final report. (The OTA is the scientific arm of Congress that lawmakers consult when they write legislation to regulate science and industry.) The general industry consensus on this provisional report was that it would be a disaster for the development of biotechnology. It called for an overregulation of the industry and required too much federal oversight of each biotechnology product that might be released into the environment. However, according to Gretchen Kolsrud, the director of OTA's biotechnology project, the document had been reviewed by a number of scientists from the industry who found nothing seriously objectionable in its content (*Feedstuffs*, 19 October, 1987).

In the mid-1980s, regulations that would require safety testing of new genetically engineered organisms were proposed by the USDA and EPA. Both agencies had considerable industry support because the biotech industry needed a government oversight program that could help build public confidence in its products. However, neither these regulations nor the provisional OTA report got anywhere. They ended up in bureaucratic gridlock under the authority of the BSCC. According to two government watchdog organizations, the major issue that the BSCC could not resolve was to what degree the EPA should regulate industry. Working through the Office of Management and Budget (OMB), some members of BSCC prevented the EPA from issuing new rules.

Although the BSCC no longer exists, other obstacles to effective regulation have taken its place:

> The BSCC met its demise in 1990. Since then, the Quayle Council [the Council on Competitiveness] has taken up the biotechnology issue, establishing a working group to oversee all related federal

policy. In February, the Council produced a slick four-color brochure, which proposes four criteria for regulating biotechnology that would severely limit earlier EPA proposals. These criteria echo the Council's overall approach to regulation—new products are generally to be considered safe until proven dangerous, instead of the other way around. For example, the Council recommends that regulation focus on "the characteristics and risks of the biotechnology product—not the process by which it is created"—which asserts that biotechnology products do not deserve special treatment under regulation. Furthermore, the Council recommends that industry be held to "performance-based standards," not "design-based standards." This means that a company need only prove that it has achieved a specific goal—for example, a requirement to contain a new organism within a certain area—and not comply with any set methods for achieving that goal.

The Quayle Council's Biotechnology Working Group, working with OMB, has taken over the role once played by the BSCC—oversight of all biotech regulations. Only now the hoops biotech regulation must jump through have moved. When the agencies first began working jointly in the early 1980's in order to coordinate biotech regulation, the goal was to apply existing regulation to new biotech programs. Laments National Wildlife Federation's Margaret Mellon (National Biotechnology Policy Center Director), "In 1984, the attitude was 'we will use existing rules to regulate [biotech].' In 1991, they are using the veto power of OMB to prevent any regulation from being approved."

By publishing new principles for biotechnology regulation, the Quayle Council has thrown the EPA and the USDA back to square one.[7]

---

[7]   Christine Triano and Nancy Watzman, *All the Vice President's Men: How the Quayle Council on Competitiveness Secretly Undermines Health, Safety and Environmental Programs* (Washington, DC: OMB Watch and Public Citizen's Congress Watch, September 1991).

The ironic outcome of all this effort to deregulate the biotechnology industry is that the Council on Competitiveness has actually stalled U.S. industries' progress. Without a sound regulatory structure, there is no security or certainty. Biotech companies are reluctant to invest in developing and marketing new products that might suddenly be subject to strict government regulation.

In February 1991, the Council on Competitiveness made public its new guidelines, which affect primarily the agricultural sector.[8] The guidelines say, "federal agencies should regulate genetically engineered products the same way that they regulate products made conventionally." The Council argues that assessments made before approving the release of a product into the environment should focus on the characteristics and risks of the product and not the *process* by which it was created. "Performance-based standards" rather than design standards should also be considered.

By deemphasizing the process involved in the creation of such products, these proposed regulations help keep genetically engineered life forms in the same category as conventionally made products, such as chemical pesticides and fertilizers, which is patently absurd.

In conclusion, legislative efforts in the United States to ensure the safe and appropriate applications of biotechnology in industry, agriculture, and medicine have reflected a growing consensus of concern. This concern is based on the lack of coordination and the jurisdictional confusion between various federal agencies involved in regulating biotechnology. Clearly the groundwork is not being well

---

[8]   S. Sugawara, "Bush Urges Less Biotech Regulation." *The Washington Post*, 25 February 1991, pp. D1, D11. For an update on proposed U.S. policy, see J.L. Fox, Biotech Regulatory 'Scope' Set. *Bio/ Technology* 10:358 1992; and R. Eisner, Whitehouse Says Recombinant Organisms Pose No Unusual Risk to the Environment. Genetic Engineering News, March 15, 1992 p. 1.

laid for the age of biotechnology, which may be short-lived if the long-term economic costs and consequences greatly outweigh the potential benefits. This may come to pass if the new technology is not applied with wisdom and foresight, the lack of which no amount of government oversight can compensate for.

A Haven Abroad?

In some countries, regulations governing the release of genetically engineered life forms into the environment are less stringent or nonexistent. This has encouraged U.S. biotech companies to conduct tests in foreign countries. In recognition of this potentially serious problem, scientists at the First International Conference on the Release of Genetically Engineered Microorganisms, meeting in the spring of 1988 in Cardiff, Wales, called for international regulations.[9] A highlight of this conference was the recent disclosure of what many critics saw as an illegal and unethical field test of a genetically engineered rabies vaccine in cattle in Argentina,[10] in which twenty cows were injected with the rabies virus spliced onto a live-virus (small pox) carrier. The test grew out of an exclusive agreement between the Wistar Institute of Philadelphia and the Pan American Health Organization: The Argentine government was not informed of the experiment. As it turned out, seventeen farm workers caring for the animals and several nonvaccinated cattle developed rabies antibodies within six months after the study

[9] See Bernard Dixon, "Genetic Engineers Call for Regulation," *The Scientist*, 2 May 1988, p. 2.
[10] See Steve Connor, "Argentinian Scandal Prompts New Gene Rules," *New Scientist*, 14 April 1988, p. 18.

was initiated, indicating that the live-virus vaccine was contagious.

This recent evidence that genetically engineered live vaccines can be transmitted from vaccinated species to others is cause for concern. Still unresolved is whether U.S. companies and researchers should comply with federal ethical and safety standards in such experiments conducted abroad.

Containment

This scenario raises the question of effective containment of all genetically engineered live vaccines. Gene-splicing techniques often employ live viruses as carriers for foreign genes (like the retroviruses now being used in poultry and the live pox virus spliced with a segment of rabies virus). Such carrier viruses could be shed by recently vaccinated animals, putting humans and other animals (especially wild birds and other creatures around farms) at risk. Genetically engineered live vaccines could behave very differently in the wide range of nontarget species upon which no safety tests have been conducted (for obvious practical and financial reasons). It is now the task of the USDA to assure containment and minimize the risks of such vaccines.

In December 1987 it was reported that scientists at the National Institutes of Health (NIH) were breeding mice that carry a copy of the genes for the AIDS virus and transmit it to their offspring (*Washington Post*, 7 December, 1987). This is reportedly the first time that the complete genetic code for an organism causing a lethal disease in humans has been introduced into another animal species. According to Dr. Dinah Singer, chairperson of the NIH's Institutional Biosafety Committee, although these mice are kept under maximal containment, there are no federal guidelines governing this kind of research. It is rumored that other institutions are planning similar experiments even though few laboratories have maximum confinement facilities. The purpose of this research is to study how viral genes are activated and to test drugs that might block ill effects.

Dr. Sheldon Krimsky of Tufts University and founder of the Council for Responsible Genetics[11] stated, "When you begin taking a virus and putting it in a different species, you can't guarantee that the methods of spread and infection will always be the same as they have been." The AIDS virus could, for example, mutate and become transmissible through urine or saliva. Any accidental escape or deliberate release of such an organism could have catastrophic consequences.

Proprietary Interests and Academia

Major obstacles to effective regulation are corporate interests and trade secrets. Biotech corporations do not want to divulge the nature of their research and development until patent protection is secured. In addition, there is the analogous veil of the industrial-military complex called "national security." It is rumored that the Pentagon is planning to use genetic engineering technology to develop lethal germs, which could have devastating consequences if it fell into the wrong hands.

Trade secrets and other proprietary and vested interests notwithstanding, private industry has an obligation to be publicly responsible and responsive. After all, the public

[11]   See the excellent publication *Genewatch*, published by the Committee for Responsible Genetics, 186A South St., Boston, MA 02111.

underwrote most of the basic research done in universities and government laboratories that has led to the industrial development and application of genetic engineering. But now private industry and state and private universities are drawing up collaborative research contracts that could lead to unprecedented restrictions on academic freedom. These restrictions would, in essence, prohibit the exchange of scientific inquiry and the advancement of knowledge. Thus, in spite of the short-term benefits of financial infusions from private industry, academia and the public interest may be shortchanged. Furthermore, land-grant colleges engaging in collaborative projects with private industry may be in violation of federal law, since they were established with public funds to research, develop, and implement advances in agriculture to help family farms and rural communities. Giant pigs and bacterial pesticides are hardly the kind of advances that will help beleaguered farming communities today.

In a critical article on industrial support of university research in biotechnology, D. Blumenthal and coauthors concluded that:

> Biotechnology companies invested about $120 million in university biotechnology research laboratories in 1984, about 20 percent of the total dollar investment that year; this contrasts with the contribution (about three to four percent annually) of industries to other types of university research. [We] analyzed the relations of 106 biotechnology companies with university laboratories. Nearly half the companies supported university research in 1984. Benefits to the companies were seen in the abundance of patent applications (two to four times as many applications per dollar invested in university as in in-house research) and trade secrets generated, which may translate into financial gain. Most arrangements were for two years or less and could have the long-range effect of shifting some

university research from basic toward applied studies. Relations of university researchers with firms and with government sponsors differ. The generation of trade secrets may jeopardize a fundamental scientific and academic value, the free exchange of information. It is considered desirable that the government, which is the major sponsor of biotechnology research in universities, remain so.[12]

Another concern expressed by Dr. Ruth L. Kirschstein, Director of the National Institute of General Medical Sciences, is that the explosive growth of commercial biotechnology now threatens to deplete universities of scientists to train the next generation of biotechnologists. She opines: "The supply of scientists who possess general knowledge in molecular biology, molecular genetics, chemistry, virology, immunology and biophysics is being rapidly outstripped by the needs of industry and academia."

Biotech industry analysts Sheldon Krimsley, James G. Ennis, and Robert Weisman have concluded that:

> The percentage of dual-affiliated faculty members at leading universities was just as surprising as the absolute numbers . . . [N]early one-third of MIT's biology department consisted of scientists with formal ties to biotechnology companies; 20 percent of the faculty in six Harvard departments were found to have similar ties.[13]

Noting the close linkage between universities and biotechnology companies, Krimsky contends that:

---

12   D. Blumenthal et al., "Industrial Support of University Research in Biotechnology." *Science* 231 (1986): 242–246.

13   *GeneWatch* vol. 7, nos. 4–5 (November 1991), p. 2.

The greatest loss to society is the disappearance of a critical mass of elite, independent, and commercially unaffiliated scientists to whom we turn for vision and guidance when we are confounded by technological choices. Once the erosion of an independent university sector is accomplished, the stage is set for what University of Washington Professor Philip Bereano aptly described as "the loss of capacity for social criticism."[14]

Public Attitudes

A Harris poll conducted in the fall of 1986 for the Office of Technology Assessment found that the majority of Americans believe that the benefits promised by biotechnology in agriculture and human medicine outweigh the risks.[15] This survey, published by the OTA as *New Developments in Biotechnology: Public Perceptions of Biotechnology*, 1987, included the following statistics:

- 62 percent of the public surveyed thought that the benefits outweighed risks.
- 55 percent of Americans are willing to chance isolated ecological damage, such as the extinction of individual plant or fish species, at risk levels of 1 in 1,000, so long as the potential risks are known.
- When the risks are unknown but the harmful consequences are considered to be "very remote," public approval dropped to 45 percent.

[14] In *Biotechnics and Society* (New York: Praeger, 1991), p. 79.
[15] Although 31 percent of Americans surveyed opposed genetic engineering on religious grounds and 35 percent felt that we should not tamper with nature, a surprising 67 percent of the Japanese public said no to research that could lead to new forms of plant or animal life (*Science*, 15 April 1988, p. 277).

- Some 63 percent of the population surveyed admitted that they knew relatively little or almost nothing about biotechnology. However, 42 percent of respondents opposed the widespread use of genetically altered microbes for agricultural purposes.
- With respect to using recombinant DNA methods to produce hybrid plants and animals, the majority of people (68 percent) were not opposed. Twenty-four percent of the people surveyed who opposed such alterations of plants and animals based their opposition on moral grounds.
- Interestingly, 42 percent of the sample contended that altering human genes to combat disease is morally wrong; 52 percent were in favor. However, when questioned about using biotechnology to prevent a child from inheriting a birth defect, 77 percent said that this was acceptable. This change in attitude may be explained by the fact that 37 percent of the participants indicated that there were genetic problems in their families. A surprising 44 percent approved of scientists altering the cells of adults to improve the intelligence or physical attributes of their children.
- With respect to safety controls, only 13 percent of those polled were willing to allow a company to make the final decision on the suitability of large-scale applications of genetically engineered organisms.

The need for regulation by federal and state government agencies and outside scientific authorities was generally recognized by the public. They felt that university scientists could be trusted most, followed by public health officials, and then federal government agencies. Significantly, when federal officials and environmental groups clash over safety issues, the OTA report concluded that, according to the data gathered in the Harris poll, the majority of Americans were more likely to believe the environmentalists.

More than half of those interviewed thought that genetically engineered products would cause a serious danger to people or the environment sometime in the future—the major risk being perceived as the use of gene-altered microbes in agriculture.

In a recent poll in Europe:

> Fewer than half thought biotechnology research on farm animals "to make them resistant to disease, or grow faster" should be encouraged. A third thought applying biotechnology to animals 'to develop life-saving drugs or study human diseases" was morally acceptable, "provided the animals' welfare is safe-guarded," but 20 percent said it was morally wrong, and 27 percent said government should decide each case. Only 13 percent thought such work justified "some animal suffering."[16]

A recent survey in Holland found that consumers "are very unhappy about eating meat from genetically engineered animals. They are either afraid it will harm them or worried about it on ethical grounds."[17]

The public is obviously not totally trusting of science and is as skeptical of the promises of medicine, agribusiness, and other industries as it is of those of politicians and government agencies. This lack of trust and healthy skepticism cannot be dismissed as superstition or ignorance. The public is assailed daily with the catastrophic failures of chemical-addicted agriculture, drug-dependent

---

[16]   D. Mackenzie, "People's Poll Shows Confusion Over Biotechnology," *New Scientist*, 13 July 1991, p. 14.
[17]   *New Scientist*, 17 August 1991, p. 9.

medicine, and other ill-conceived scientific and industrial "advances."

Could these catastrophes have been avoided if there had been more foresight? The shortsightedness of private interests and the consistent lack of concern for the public good and related long-term social, ethical, and environmental consequences are due in part to an almost total lack of public involvement in policy-making at all levels of government, industry, education, and research.

The hallmark of a technocratic society is the appropriation of the public right of involvement and ultimately of democratic self-determination. And when children are raised to obey the voice of authoritarian rationalism without question despite what they feel and know intuitively, the viability of technocracy is assured—at least for a few generations, until either the human spirit rises up in revolt or the system disintegrates under the combined weight of unmanageable complexity and bureaucratic-managerial entropy. Hence, public involvement in the policy decision making of the biotechnology industry is important at all levels, both nationally and internationally.

Because the general public is being led by the biotech industry to believe that biotechnology is a panacea for many of life's problems, it is difficult to take the position of a critic. Such a position seems both antiprogress and antisocial. To express concern for animals' welfare or nature's creation is seen as relegating human interests to the backseat. It is easier for the industry to gain public support by arousing hope and faith in science than for its critics to appeal to reason and evoke public fear and doubt. No one wants to bear the burdens of fear and concern for the adverse consequences of biotechnology. And to believe that the industry can regulate itself and the government can ensure safe, humane, and environmentally neutral (if not enhancing) applications of biotechnology is simply wishful thinking. We have perhaps less to fear of this new technology than we have of the human psyche that protects itself through rationalization and denial: The potential risks and adverse consequences of biotechnology are ignored, and to voice legitimate concern is to be not a realist but an alarmist.

Because public understanding of genetic engineering and biotechnology is limited, its complexity intimidating, and its consequences seemingly far removed from the community and our daily affairs, it is difficult to arouse enough public concern to get people to participate in the processes that lead to the ultimate adoption or rejection of technological innovations. However, the nascent biotechnology industry has already awakened considerable public concern in its eagerness to get new products tested and marketed. Some biotechnology companies have even sought liability insurance, just in case of a major catastrophe.

Several serious infringements of government safety guidelines have made news headlines. In April 1986, the EPA suspended the test permit and fined Advanced Genetic Sciences of Oakland, California, for secretly testing a frost-resistant strain of *Pseudomonas syringae* in the open environment by injecting it into trees on the company's rooftop. But a month later, the EPA approved a permit allowing researchers from the University of California to spray these genetically engineered bacteria on test fields of potatoes. One of the researchers involved was Dr. Steven E. Lindow, who had worked with Advanced Genetic Sciences.

In the spring of 1986, Biologics Corporation of Omaha had its product license temporarily suspended by the USDA after an employee reported that the company had injected pigs with an anti-pseudorabies live-virus vaccine without following established government safety guidelines. It was later found that the USDA had ignored its own guidelines by granting this company a license to distribute the vaccine for commercial use in January. The

USDA released a thirty-page environmental-impact study and lifted the suspension after two weeks.

On 13 June 1987, Gary Stroble, a professor of plant pathology at Montana State University, released genetically engineered *Pseudomonas syringae* as an act of civil disobedience. He injected the bacteria into fourteen elm trees, hoping that the bacteria would release an antifungal substance to stop Dutch elm disease. Stroble did not apply for an EPA permit until after he had injected the trees because approval would have taken too long. He did not want to lose a year's work by having to wait until the next season to inject the trees (*Washington Post*, 14 August, 1987). Clearly the narrowly focused altruism of some scientists borders on obsessiveness and is another reason for concern about the expanding field of biotechnology.

Ironically, around this same time the National Academy of Sciences released a report that concluded that there "is no evidence that unique hazards exist either in the use of [recombinant DNA] techniques or in the movement of genes between unrelated organisms."[18] According to panel chairman Arthur Kelman, professor of plant pathology and bacteriology at the University of Wisconsin, the main purpose of the report was to reassure the public that benign organisms would not be converted to threatening pathogens, based on the available biological evidence (*Washington Post*, 19 August 1987). This obfuscation of the potential ecological risks of releasing genetically engineered organisms into the environment is blatantly obvious. The real concern, which the panel sidestepped, is not simply that organisms may be converted to threatening pathogens, but that serious ecological imbalances might

[18]   National Research Council, National Academy of Sciences, *Introduction of Recombinant DNA-Engineered Organisms Into the Environment: Key Issues* (Washington, DC: National Academy Press, 1987).

occur following the release of genetically altered bacteria and other organisms into the environment.

The majority of scientists who serve the biotechnocracy operate on the arrogant and paternalistic assumption that science knows best. Nature isn't perfect, their reasoning goes, but it can be improved upon. Likewise, science isn't perfect, but any problems that might arise from the life sciences' meddlings with nature can be easily fixed. They like to quote Sophocles: "One must learn by doing the thing, for although you think you know it, you may have no certainty until you try." But this attitude is cause for alarm for those who do not share the hubris of scientism. As philosopher and farmer Frederick Kirschenmann said, "We can never see things the way they are in themselves, we can only see them through the reflection of our own perceptions." So much for scientific objectivity. As naturalist-ecologist John Muir once advised, "We can never do just one thing."

Lawrence Bush and coauthors of the book *Plants, Power and Profit* conclude that, "neither biotechnologists nor policy makers are adequately equipped to make choices about which is the preferable future."[19]

James Watson, who won the Nobel Prize for his work on the structure of DNA, was clearly disturbed about nonscientists becoming involved in the debate over the ethics, safety, and research guidelines of genetic engineering that were adopted at a conference in Asilomar in 1975. His arrogance and elitist attitude are expressed in a statement he made:

> Although some fringe groups ... thought this was a matter to be debated by all and sundry, it was never the intention of those who might be called the molecular biology establishment to take the

[19]   L. Bush et al., *Plants, Power and Profit* (Oxford: Blackwell, 1991), p. 237.

issue to the general public to decide. We did not want our experiments to be blocked by over-confident lawyers, much less by self-appointed bioethicists with no inherent knowledge of, or interest in, our work. Their decisions could only be arbitrary.[20]

The public's lessons from ozone-destroying chlorofluoro-carbons (which scientists and industry thought to be safe) and from the misapplication, inept or deficient regulation, inadequate and even fraudulent safety testing, and manufacturers' concealment of health hazards of chemical pesticides should provide sufficient pressure to ensure that bacterial pesticides and other genetically engineered life forms will never be deliberately released into the environment unless their release can be regulated effectively and all potential risks have been identified. As former EPA chief William Ruckelshaus stated at the 1985 National Academy of Sciences symposium on biotechnology, "This is the last chance we have to do it right the first time."

Clearly one of the greatest risks in agricultural biotechnology is conceptual. The inherent risks of ideologically unsound concepts such as giant pigs and herbicide-resistant crops must be recognized, since their application is not in the best interests of agriculture, but in the best interests of monopolistic private enterprise. The long-term social and economic consequences to the farming community need to be considered, as well as the obvious ecological and animal-rights and -welfare concerns.

Before adequate national and international regulations for the biotechnology industry can ever be formulated and effectively adopted, a much more sophisticated grasp of the industry's long-term influences on agriculture, society, and the natural environment is needed. These are explored in the next chapter.

---

[20]   J. D. Watson and J. Tooze, *The DNA Story* (San Francisco: W. H. Freeman, 1981), p. 49.

CHAPTER 4

GENETIC ENGINEERING, AGRICULTURE, AND SOCIETY

A technological assault is being prepared that will transform the economies of developed and developing nations. Its substance is the engineering of life processes for commercial ends.
—EDWARD YOXEN, *THE GENE BUSINESS*

*T*here is nothing miraculous about our new-found ability to engineer animals, plants, and bacteria genetically. This has been going on in nature since the beginning of life. The process is called evolution.

But we humans think we know better and presume to use our power over the gene to control the direction of evolution and "improve" upon nature. Actually, we are trying to improve conditions for ourselves too often at nature's expense; the environment is perceived as a resource, and all nonhuman life is inferior and thus expendable or exploitable without any ethical constraints. A case in point is the wholly self-serving creation of transgenic plant and animal species. The ethical aspects of switching genes between species to make them more useful to us were never considered, nor were the long-term ecological and other consequences envisioned.

We must stop to consider the hazards and consequences of misusing our power over nature. Some of the consequences we are already aware of—floods, droughts, salinization, radiation, pollution, and thermonuclear accidents. We are now predictably engaged in an international biotechnological race with Japan and the European Economic Community.

Agribiotechnology

Agribusiness is engaged in a headlong rush to increase productivity of animals and crops and to stop diseases that other production-boosting manipulations have caused. The new biotechnology paradigm follows the old one of trying to make farm animals and plants more productive and efficient through selective breeding, special feed and fertilizer, hormones, and so on, and then correcting the resulting production-related diseases, notably increased susceptibility to infectious microorganisms, pests, and parasites. This circular approach necessitates more research and technological "fixes," such as drugs, vaccines, and poisonous pesticides. These additional problems require even more governmental oversight and costly regulatory bureaucracies to protect consumers and the environment. They also mean more profits for the manufacturers thereof.

Biotechnocrats dream of creating animals and crops that are highly fertile, productive, and disease resistant. The reality of the agricultural system as it now exists is that productivity and fitness—health—are inversely correlated. Can genetic engineering change this so that animal and plant health is not jeopardized by transgenic and other manipulations? Will a highly controlled technosphere be as productive in the long run as a less intensively controlled and more natural biosphere? Predictably not. Increased production through genetic engineering could exhaust nonrenewable resources more rapidly and fail to feed a larger and more dependent human population. This is already borne out by the almost total failure of the Green Revolution. Agribusiness technology experts, super breeds of crops, irrigation and hydroelectric dams, chemical fertilizers, pesticides, and agripoisons exported to less developed countries produced great short-term profits but destroyed already existing, more regenerative, traditional farming practices, ultimately destroying the communities and the fragile land.[1]

Genetically engineering crops that are drought and salt resistant may be more prudent and appropriate applications of biotechnology than for companies to develop bacterial pesticides and seeds that are resistant to their own herbicides. But since drought and salinization are mainly man-made, the more prudent course would be to correct these problems rather than use genetic engineering as a new technological fix. Furthermore, natural deserts, swamps, and salt marshes need to be preserved to protect biodiversity and the integrity of the Earth's ecology. This means that developing genetically engineered

---

[1] Prior to this period, circa 1960–1980, great and probably irreparable harm had already been done to the traditional, often socially just, and sustainable farming practices of third-world tribal communities by white colonialists, many of whom ran vast plantations, cattle ranches, and mining and timber enterprises, especially in Africa and South and Central America.

plants that could invade these ecosystems, or be deliberately planted for commercial gain, should be strictly regulated so as to minimize further loss of natural ecosystems. It is unlikely that genetic engineering technology will be appropriately and reliably applied as long as it remains within the narrow, "agricidal" paradigm of multinational agribusiness.

Many of the promises of agribiotechnology will never come to fruition unless innovations are carefully integrated with sustainable agriculture. Nicamor Perlas writes:

> But what of the promised benefits of genetically engineered crops? So far, there is room for skepticism about at least some of the claims of the biotech industry researchers. The hope of increasing the protein content of plants may be more wishful thinking than practical agriculture. Genetic engineers seem to have forgotten the hard lessons of the high-lysine corn panacea of the late 1960s and early 1970s. When scientists at Purdue University discovered a corn variety high in the amino acid lysine, there was general euphoria over the prospect of vanquishing malnutrition in countries with corn-based food systems. When they tried to apply their discovery, however, scientists quickly learned that the new high-lysine corn varieties had a smaller yield than conventional, lower-protein corn types. In addition, the high-lysine corn had more insect problems and was of poorer quality when milled and processed. Eventually the seed companies gave up trying to market high-lysine corn.
>
> Another promised benefit is the transfer of nitrogen-fixing genes from bacteria to cereal grains. If scientists succeed in finding the key to breaking into the labyrinth of interacting genes that govern nitrogen-fixing, they are still faced with the old problem of energy trade-offs. Nitrogen fixation, the process by which nitrogen in the air is taken up by soil bacteria and made available for plants, is an energy-intensive process that drains

the carbohydrate reserves of plants. In all likeli-
hood, transferring the nitrogen-fixing gene from
bacteria to cereals will reduce the yield.

Even were genetic engineers to succeed in improving crops as intended, it is doubtful that the
problem of hunger would have been solved. Many
food analysts, including economists at the World
Bank, are belatedly recognizing that hunger is
a distributional and political problem. The earth
produces enough food to feed all the hungry people
in the world today, but economic and political factors prevent the flow of food to the needy.[2]

One goal of agribiotechnology is to enhance the bio-
competitiveness and adaptability of crops (including trees)
by enhancing plant resistance to drought, salinity, disease,
pests, and herbicides. Another is to link biocompetitive-
ness and adaptability with bioproductivity, using genetic
engineering to enhance growth, productivity, nutrient
value, and chemical composition (for the biosynthesis of
pharmaceuticals and industrial chemicals). The selective
breeding, cultivation, and genetic engineering of algae
is another branch of agribiotechnology that poses fewer
environmental risks but is still in its infancy. A few companies are also expanding on the principles of biological
pest control (as distinct from chemical pest control) by
selectively breeding and genetically altering various microorganisms that kill plant pests such as the Colorado beetle
and corn root worm. Another branch of food production—
aquaculture and mariculture—is applying biotechnology
to enhance the growth and disease resistance of trout,
catfish, carp, salmon, shrimp, and other species. Some fish
species may also be used in "molecular" farming to produce
useful protein and pharmaceuticals.

Dr. David Wheat, agricultural consultant for Arthur D.
Little Company, in an address before the 1988 Agribiotech

---

[2]   "The Sustainable Agriculture Movement," *Orion Nature Quarterly*,
Spring 1988: 35–47.

International Conference and Exposition (Washington, D.C., 26 January 1988), stated: "Biotechnology will have significant competitive impact in agribusiness but most of the benefits from using these new techniques will flow to larger firms." He predicted that agribiotech will produce few new products but will become a key technology in existing businesses, such as seed, crop chemicals, and animal health. Large companies will have to acquire biotechnology in order to compete effectively, and small firms are likely to be forced to sell their technology or pursue specialty or niche markets. Significantly, according to Dr. Wheat's analysis, those companies with a strong monopoly on the seed industry, notably Pioneer, Sandoz, and to a lesser degree Imperial Chemical Industries, Dow, and Ciba-Geigy, have a distinct advantage over other companies with little or no control of the seed industry.

Other companies such as Monsanto, Upjohn, Elanco, and Pitman-Moore are investing tens of billions of dollars in research and development of genetically engineered porcine and bovine growth hormones to boost the growth of pigs and beef cattle and the milk yield of dairy cows, which they anticipate will reap great profits worldwide.

A 1987 survey conducted by the University of Wisconsin, Madison, of Wisconsin dairy farmers' attitudes toward bovine growth hormone (BGH) (*AgBiotechnology News*, July/August 1987) revealed that only 11 percent favored private industry's effort to develop and market this hormone, 56 percent were opposed, and 33 percent were neutral. Of the 270 dairy farmers surveyed, 22 percent flatly opposed the university conducting any research on BGH.[3]

---

[3]   In late 1991 there were reports of university and corporate coverups of dairy cow health problems arising from injections of BGH. Problems included increased incidences of disease, higher mortality and infertility rates, and birth defects. The influence of corporate interests seeking FDA approval for the harmful product was so strong that one university was accused of burying the deformed calves of BGH cows.

In a poll of 1,900 Iowa farmers by Iowa State University (*AgBiotechnology News*, July/August 1987), it was revealed that farmers see biotechnology as a mixed blessing. Eighty-two percent felt that biotechnology was desirable if it enabled farmers to become less dependent on chemical pesticides. Sixty-four percent saw it as undesirable if it benefited large farm operations more than small and mid-sized farms, and 71 percent agreed that an adverse consequence of biotechnology would be farmers becoming more dependent on large corporations for such things as seeds, feed additives, and growth hormones.

Although increased food production was a major concern of many of the farmers polled, 45 percent viewed increased food production as undesirable and 46 percent viewed doubling milk production as undesirable if it involved embryo transfer, gene splicing, or hormone treatments. A Jones County farmer wrote, "Biotechnology needs to solve problems of storage and distribution of food supplies to hungry people and not just . . . result in greater production." "Raising more is sure not the answer," another Iowa farmer observed.

Billions of dollars are being invested in applying biotechnology to increase farm animal production and disease resistance, even though no cost analysis has been done to compare this approach with potentially less costly alternatives such as improved traditional breeding, management, and feeding practices. I know of no studies that have considered the long-term environmental and economic consequences of using biotechnology to maintain or expand meat- and milk-based agricultural practices.

The most effective, if not enlightened, public response other than vegetarianism is to support consumer-interest organizations that work with government to instigate appropriate labeling of all genetically engineered food products (including milk from cows receiving BGH). Natural and organically raised foods should also be labeled so consumers can support the producers and retailers of such foods. (I do not see fish and other marine and aquaculture animals such as shrimp and oysters as acceptable alternatives to meat. Ocean food sources are seriously depleted and polluted, and aquaculture products require drugs to control disease and boost productivity as well as costly nutrients in their feed, as do farm animals.)

It is surely bad medicine to prescribe yet another capital-intensive innovation in the form of biotechnology for an already overcapitalized and overproductive agricultural industry. The proponents of biotechnology consistently downplay its potential risks and costs while making unfounded promises. One great promise is that biotechnology will boost agricultural productivity and make it more efficient, thereby reducing costs so that the United States will once again have a competitive edge in the world market and ensuring that farmers will be able to grow more crops for export. The truth remains, however, that it is imprudent to adopt any innovation without a thorough and continuing assessment of its impact on both the natural environment and the social environment, such as rural communities and family farms. Otherwise, biotechnology will be inappropriately applied to the detriment of both.

One of the most poignant and tragic consequences of the technocratic ideology, which places the values of efficiency and productivity over all else, has been the demise of the family farm and of rural life. Cultural values, implicit in the concept of agri*culture*, have been superseded by those of high-tech agri*business*, where human interests —specifically concern for the viability of the family farm structure and rural communities—have been sacrificed. Certainly poor business acumen, greed, and encouragement by local banks to go too deeply into debt have contributed to the demise of many farmers. But economic survival is also equated with having to adopt capital-intensive technologies that are almost invariably inhumane to farm animals, ecologically unsound, and entail the use of chemicals and drugs harmful to the environment, consumers, and wildlife, yet give rewarding short-term profits. They also provide a competitive edge over other farmers who either cannot afford or are ethically opposed to such innovations.

What is needed is a balance between the technocratic values of agribusiness and the ethical principles of a humane, sustainable (ecologically sound), and democratic agriculture. Regrettably, the latter is regarded as a romantic regression and the former as progressive innovation. In reality, true agricultural progress lies in the integration of the two.

A special report published by the Office of Technology Assessment (OTA) in March 1985 emphasized that new technology and genetic engineering methods will greatly increase the trend toward larger farms over the next fifteen years and urged the government to provide more help for mid-sized farms,[4] which will otherwise be forced out of business.[5] This problem is now being aggravated by the tax structure and price-support programs under the Bush administration. Significantly, in the introduction to the congressional report it was stated that:

> Before the turn of the century cattle ranches in Texas may be able to raise cattle as big as elephants. California dairy farmers may be able to control the sex of calves and increase milk production by more than ten percent without increasing

[4] In the U.K., a law has already been passed to protect traditional family farms under the Countryside Preservation Act. Several members of the European Economic Community now provide significant subsidies to farmers who continue and convert to "natural", organic, and more humane farming practices.

[5] *Technology, Public Policy, and the Changing Structure of American Agriculture: A Special Report for the 1985 Farm Bill* (Washington, DC: U.S. Congress, 1985).

food intake. And major crops may be genetically
altered to resist pests and disease, grow in salty
soil and harsh climates, and provide their own
fertilizer.

DVM Magazine (May 1986) went on to summarize this
OTA report as follows:

> Genetic engineering and other emerging veteri-
> nary and agricultural technologies may end up
> raising farm costs and reducing the number of U.S.
> farms by half, a congressional report predicts.
>
> *Technology, Public Policy and the Changing Structure
> of American Agriculture* is a two-year study of bio-
> technology conducted by the Congressional Office
> of Technology Assessment. It is the first in-depth
> analysis of new biological, mechanical and man-
> agement technologies for farming.
>
> In the near future, for example, the average cow
> will bear more calves because dairy farmers are
> utilizing embryo transplants and advanced hered-
> itary techniques to produce superior embryos and
> then artificially transplant them in the womb.
>
> Cows also will be able to produce twice as much
> milk by the end of the century with the help of a
> genetically engineered growth hormone.
>
> According to the report, while these technologi-
> cal advances will help the largest and wealthiest
> farms make major gains in productivity and effi-
> ciency, they also could mean the end of small and
> medium-size farms which cannot afford the cost of
> such technology.
>
> By the year 2000, more than a million small and
> medium-size farms, of the nation's total of 2.2
> million, will disappear, the report says.

Some critics of agribusiness contend that the last thing
we need is more productive crops and livestock, since
overproduction is already a major problem. That genetic
engineering could enhance the productive efficiency of

crops and livestock is dubious: Trying to get more for less causes stress, resulting in lowered disease resistance and production-related diseases. Furthermore, it would be prudent for the United States to drastically reduce its overall production and consumption of farm animal produce for reasons of consumer health, the economy, and the long-term adverse ecological consequences of monoculture farming.

There has been virtually no public or intragovernmental debate on the impact of biotechnology on the public interest. Meanwhile, large petrochemical companies such as Ciba-Geigy, Atlantic Richfield, Monsanto, and Shell have acquired some eighty seed and plant science firms and have made significant progress in developing seeds that are resistant to their own herbicides. These chemical giants are now pushing to have their new plant varieties patented. Some fear that this will lead to a global monopoly of seed stock, which could be highly detrimental in terms of eliminating alternative varieties of seed stock and alternative agricultural practices.

The Supreme Court has ruled that genetically engineered microorganisms can be patented. And the appeals board of the U.S. Patent and Trademark Office has ruled that all genetically engineered plants, seeds, and tissue cultures can also be patented. According to reporter Marjorie Sun (*Science* 230[1985]:303), this will be a boon to the agricultural biotechnology industry because:

> Previously, new types of plants were only narrowly protected by two federal statutes. Under these laws, if a plant variety was genetically modified, for example, by the insertion of a specific disease-resistant gene, a plant breeder could only obtain protection for the specific variety, even though the modification could be applied to thousands of other varieties. The appeals board decision now allows a breeder to patent generically all varieties

with the change, according to patent office official Charles Van Horn.

This ruling can only further the monopolistic trend of capital-intensive, high-tech agriculture.

In a recent study entitled "The Potential Impact of Molecular Biology and Genetic Engineering in Agriculture 1982–2000" by Bernard Wolnak and Associates, a management consulting firm, the following points were raised (*Feedstuffs*, 8 July 1985):

- Biological nitrogen fixation consumes enormous amounts of energy and would have to be fueled for the most part by metabolism of carbohydrates from the plant itself. The resulting yield loss is frequently of greater economic value than the reduced expense of nitrogen fertilizer.
- Historically, the improvement of the nutritional quality of cereals has resulted in significant yield losses and other agronomic problems. Genetic engineering is not likely to reverse this trend.
- The development of highly profitable crop agriculture programs based on molecular biology and genetic engineering will be limited to those few companies that are able to make the difficult transition from impressive research feats to successful mass marketing.

Biotechnology, like other technologies, can force upon us things we do not want and would probably be better off without. A 20 to 40 percent increase in milk production in dairy cattle by using genetically engineered bovine growth hormone (somatotropin) will have a severe socioeconomic impact on the already overproductive dairy industry.[6] It could mean even greater dairy surpluses and lower market prices, which will certainly put many smaller producers out of business. And dairy cows will suffer and burn out even faster than they do now (see Chapter 5).

It is unlikely that the dairy industry will be able to outlaw the use of this hormone, even though the indirect costs to the farming community will far exceed the benefits, which will accrue almost exclusively to the investors and manufacturers. Recent research has shown that cows subjected to this biotechnology are more productive, but they also require more feed.[7] Thus the animals are not significantly more efficient. Untreated dairy cattle are already being pushed beyond their productive limits, which results in suffering and a variety of so-called production diseases. These problems and the suffering associated with them are likely to increase by treating high-yield milk cows with growth hormones to make them even more productive.

The increased efficiency or cost savings comes from having to house fewer animals—forty growth-hormone-treated cows produce the same amount of milk as fifty or even a hundred untreated ones. However, treated cows are likely to be culled at two to three years of age, instead of four or five (which is the age that today's dairy cows are spent). So with a faster turnover and the costs of raising relatively

---

[6] R. J. Kalter, "The New Biotech Agriculture: Unforeseen Economic Consequences," *Issues in Science and Technology*, Fall 1985: 125–133.

[7] The U.S. government has given hundreds of millions of dollars to compensate the beef industry, harmed by the 1986 dairy buy-out program when a glut of hamburger meat from surplus dairy cows saturated the market. In 1991, almost one million tons of beef were held in cold storage at considerable public cost due to overproduction in the industrial democracies of the world—which together subsidize intensive animal agriculture to the tune of $100–$150 billion a year!

more calves to replace burned-out cows, this growth hormone treatment will be of little economic benefit to most dairy farmers. This could also mean fewer calves for the veal industry and more hamburger meat for the burger industry from spent dairy cows, which would have an adverse impact on beef cattle farmers.[8] Aside from these adverse consequences, many dairy farmers would be forced out of business. The increased efficiency of this biotechnology would be realized only by well-established large "super farms" that already have an economic advantage over smaller dairy operations.

Environmental Concerns

In 1986, American companies and the government spent an estimated $2 billion in biotech research. According to *Genetic Engineering News*, 114 of 322 companies responding to its survey were agricultural products companies.

Monsanto is one of a handful of companies developing bacterial pesticides. One technique is to insert the gene that produces the poison delta endotoxin from *Bacillus thuringiensis* into the bacteria that naturally colonize the roots of plants such as corn. Root-eating pests would then be killed, but beneficial insects could also be killed. The risk of the bacteria transmitting engineered genes to other bacterial species in the soil is also very real.

In March 1988 the Environmental Protection Agency (EPA) gave a tentative go-ahead for field tests of corn seedlings injected with genetically engineered bacteria that

---

[8]   R. J. Kalter et al., *Biotechnology and the Dairy Industry: Production Costs and Commercial Potential of the Bovine Growth Hormone*, A.E. Research 84-22 (Ithaca, NY: Cornell University Department of Agricultural Economics, December 1984).

produce a toxin lethal to the corn borer. Opponents to this field trial pointed out that the microbe could spread to other plants, making weeds resistant to insect predators. However, because the bacterial toxin is not specific for the corn borer, it could kill beneficial insects as well. Other insect species could become resistant to the toxin, and the ecological perturbations that ensue could result in the evolution of new crop pests.

According to *Biotechnology News Watch* (21 March 1988), biotechnologists at the French National Institute for Agronomic Research, in collaboration with researchers at the University of California at Riverside, are exploring the possibility of making certain useful insects resistant to pesticides. The tentative plans are to develop such resistance in honeybees, ladybugs, and parasitic wasps. This would be yet another step in the wrong direction toward a reduction in natural biological diversity and is the kind of thinking that gives the biotechnology industry a bad reputation among conservationists and ecologists.

It was once believed that agrichemical pesticides and herbicides would rapidly disintegrate and lose their toxicity. We have since learned otherwise. Likewise it is believed that the new bacterial pesticides will die quickly after they have done their work, such as killing pests with the toxins they produce or inhibiting frost from forming on crops. Although this may or may not be wishful thinking (some bacteria in the roots may not migrate far, and some could be engineered with "suicide genes" so that they are short-lived), it is extremely narrow thinking, because nothing in nature acts in isolation. These bacteria could cause long-term ecological perturbations because of their influence on other species of bacteria and other living organisms.

These catastrophic probabilities aside, we should ask whether we need these new living pesticides in the first place? Is this the right direction for agriculture to take? Many experts insist that it is not, since it is a continuation of non-sustainable, capital-intensive farming. What is needed is an ecologically sound regenerative agriculture that works in harmony with nature's laws. The application of genetically engineered bacteria as pesticides is simply a misuse of our power over the gene and over life itself. But this is not to imply that genetic engineering per se has no applicability to agriculture. The technique can be used to enhance plants' resistance to drought and disease and to improve crop yields and nutrient value.

Genetically engineering certain crops, such as soybeans, to resist a potent herbicide that kills everything else in the fields is another example of the misuse of our power over the gene. Although such herbicide-resistant seeds could be highly profitable to the manufacturers, this is surely the wrong direction to take. Drug-dependent farming is hazardous to all life.

It is especially disconcerting that genetic engineers believe that removing a gene from a bacterial organism or adding one does not create a new and potentially dangerous life form. Since similar mutations occur in nature, why worry about man-made mutant life forms? The real dangers lie in the potential adverse environmental impact of large numbers of mutant organisms being released and in the transmission of the newly engineered traits to other bacteria.

For example, Dr. Steven Lindow has engineered a strain of *Pseudomonas syringae*, a common bacterium on plants, so that it does not form ice crystals, which cause frost damage to crops. He has stated:

> The bacterium we are working with is neither new nor novel to California. The procedure we use to construct it is somewhat novel, but that in itself doesn't lead to a novel organism.... Such genetic engineering is not unlike what could have happened in nature by various mutational processes. (*Los Angeles Times*, 8 September 1985).

Although Dr. Lindow's reasoning is logically based on what is known about natural mutations, additional facts must be presented before we can blithely accept his analysis of the situation. The lipoprotein coatings of certain bacteria that occur on plants are blown from the plants and soil (along with similar materials from bacterially degraded vegetation) into the atmosphere. Once in the upper atmospheric regions, these particles act as nuclei around which water collects and freezes to form ice. Some scientists consider this process absolutely essential for rain to fall (*Science News* 127[1985]: 282). Genetically engineered strains of *Pseudomonas syringae* that prevent ice formation on crops could conceivably cause serious climatic perturbations that inhibit rainfall and cause drought. Ecologist Eugene P. Odum cautions:

> Since microorganisms play major roles in maintaining earth's life-support systems, we need to be especially careful about tinkering with decomposition and other recycling processes. Unlike the life-support system of a spacecraft, which is mechanical and man-made, the biosphere is bioregenerative and self-regulating. Since we did not build it we don't know much about how it really works, and we have shown little interest in studying it at the necessary large scale until recently, when malfunctions have begun to appear due to human impacts. The case of the ice-nucleating bacteria is an excellent example of the need for a more holistic assessment that allows for consideration of roles and functions other than the one that seems undesirable. (*Science* 229[1985]: 1338)

The nonselective killing of insect pests and weeds (which may harbor beneficial insects) is imprudent. Some will develop resistance, and new pest and weed problems will arise. Already resistance to *Bacillus thuringiensis*, which has been widely used as a bacterial pesticide, has been demonstrated in the Indian meal moth, a major pest of stored grain.[9] That insects and other organisms will not develop resistance to new genetically engineered bacterial pesticides is wishful thinking. Research to combat resistant strains will become necessary, and a profitable new industry will arise, adding speed to the poisonous pesticide-herbicide treadmill of today's agrichemical industry.

According to a report published in 1984 by Michael Dover of the World Resources Institute and Brian Croft of Oregon State University, the number of insects that have become resistant to pesticides almost doubled between 1970 and 1980—from 224 to 428 species.[10] For instance, in 36 different countries 25 species of mites, caterpillars, beetles, and other insects that attack cotton plants are now resistant to all available pesticides. According to this report, more than 150 kinds of fungi and bacteria are now resistant to pesticides, an increase from an estimated 20 in 1960. Furthermore, over 50 species of weeds are resistant to one or more herbicides.

Several ecologists have expressed concern over the cavalier attitude of geneticists and molecular biologists toward the release of genetically engineered organisms into the environment. R.K. Colwell et al.[11] raise the following concerns, which I have summarized as follows:

- Genetically engineered traits such as herbicide resistance, insect resistance, stress tolerance, and nitrogen-fixing ability could be transferred from crops to weeds through hybridization to produce new and more troublesome weeds. Many weeds

---

[9] William H. McGaughey, *Science* 229(1985): 143–194.

[10] *Getting Tough: Public Policy and the Management of Pesticide Resistance* (Washington, DC: World Resources Institute, 1984).

[11] R. K. Colwell et al., "Genetic Engineering in Agriculture," *Science* 229(1985): 111–112.

are known to hybridize with crop species; commercial sorghum, for example, will hybridize with Johnson grass to produce a vigorous perennial hybrid weed in sorghum fields.

- The natural homeostatic balance of diverse species within ecosystems has been thought to be a controlling factor in stopping a genetically engineered organism from explosive expansion. But this may not be true: Witness the dramatic population explosions of the fruit fly *Rhagoletis pomonella*, which infests apples, and the collared turtle dove, both of which may be mutations. Engineering organisms to be disease or drought resistant, or more hardy and able to thrive in poor soil or at high or low temperatures, may attenuate the effects of natural population-limiting factors. Thus the possibility of explosive expansion is enhanced, with potentially catastrophic long-term ecological perturbations. Genetic engineers not only lack the essential knowlege of predictive ecology, but they have no way to ensure that the bacterial and plant "neobiotes"[12] they create are not *too* viable and competitive so as to pose a serious environmental risk.

- Although genetic engineers may feel confident that the engineered traits of an organism permit an *a priori* prediction of its environmental impact, Colwell et al. emphasize that "the phenotype of any organism, including its ecological role, is not fully predictable from genotype alone. Pleiotropies, multiple phenotypic effects from the same genetic change, can be ecological as well as physiological or morphological and may include unexpected interactions between species."

---

[12] This is my neologism for genetically engineered new (neo-) life forms (-biotes):

- The now well-documented phenomenon of plasmid transfer of resistance to antibiotics in bacteria to other microorganisms raises the specter of the rapid spread of novel artificially induced genes to other organisms, the ecological consequences of which cannot be determined.

Before genetically engineered microorganisms are used to enhance disease resistance in plants and animals, they should be "agoraphobic" (will not spread outside the host plant or animal) and "suicidal" (will die after a given period of usefulness).

Certainly genetic engineering could be applied beneficially to increase the natural genetic diversity of highly inbred domestic varieties of crops. (For example, 100 percent of the banana export crop in South America in 1984 was derived from a single variety.) Greater genetic variability to enhance disease resistance and overall vigor (so-called hybrid vigor) could be accomplished by gene-splicing inbred crops with the genes of wild varieties.

Probably the most promising and as yet virtually unexplored area of genetic engineering in agriculture entails the propagation of single-cell organisms, such as algae, that have been genetically engineered to produce amino acids and other essential nutrients. Such technology would enable us to synthesize a variety of essential nutrients. We already have the technology to create from such nutrients highly palatable, texturized, and flavored substitutes for meat. As the human population increases every year, the consumption of meat as a dietary staple becomes increasingly unacceptable. But we should not be sanguine about synthetic, genetically engineered foods.

These brave-new-world engineers should reflect upon the hazards of introducing new life forms into the environment. Farmers' fields and animal confinement buildings are not totally isolated. For example, a new intestinal or rumen bacteria that rapidly breaks down lignin (wood chips) and speeds up the digestive process and yield over time of farm animals could transfer its trait to other bacteria in the environment if the manure is not contained. We could end up with no trees. The engineers could try to create very short-lived organisms, but such organisms would have to be alive for some time in order to function, and therein lies the inherent risk.

One example of the possible effects is the damage caused by the inadvertent release of killer bees in South and Central America.[13] Another was the release of *Serratia marcecens*. This bacterium, which scientists thought to be harmless, was propagated and released in the 1950s into New York subways and airports and from high-rise buildings in Los Angeles by the Central Intelligence Agency as a "model" for bio-warfare. Although it took time before its impact was felt, this "harmless" organism has killed thousands of people in hospitals from so-called nosocomial disease.[14] The organism is highly resistant to antibiotics and routine sterilization procedures and is often injected into people in hospitals when they are given infusions. They die from bacterial pneumonia and other complications.

Little wonder that more and more conservationists and humanitarians are concerned about the destructive potential of genetic engineering. We must acknowledge these hazards. The industry cannot be stopped because it is government endorsed and growth stimulated by international competition (primarily between the United States

---

[13]   The "killer" bees were developed by scientists in Brazil in 1956 by crossing honeybees with African bees. Some escaped and caused havoc in the honeybee industry and in some orchards and plantations in Central and South America. For an excellent review see Killer Bees: *The Africanized Honey Bee in the Americas* by Mark L. Winston. (Boston: Harvard University Press, 1992).

[14]   See Dr. Hugh Drummond, *Mother Jones*, September/October 1983, p. 49.

and Japan). For ecological reasons, I believe that genetic engineering of bacteria and other microorganisms should be strictly confined to P–4 laboratory conditions (maximal containment), and that they should never be released into the environment without full assurance of their zero-risk potential.

Environmentalist Jack Doyle points out:

> Only since 1983 have government agencies and scientific institutions begun to think indepth about what genetically altered organisms might do in the environment outside the laboratory. And only as recently as June 1985 did molecular biologists and ecologists, meeting in Philadelphia for the first time on the question of deliberate release and ecological risks, begin a scientific dialogue on the subject. . . .
>
> In February 1985, the Cornell University Ecosystems Research Center issued one of the first comprehensive overviews of what was possible and what was lacking in the way of assessing the environmental risks of new biotechnology products. The findings of that report are fairly astonishing and underscore how little we know about microorganisms—how they survive, why some grow and others do not, and how they disperse in the environment.
>
> But the clincher in the Cornell study is what is said about our ability to make predictions about the behavior of organisms in the environment. "Methods for predicting the likelihood of survival and proliferation of a given organism in the environment are crude," says the report. "Methods are available for assessing some potential effects, but there are many deficiencies in current knowledge and theory. Generally, we lack any true data base against which to compare test results or predict environmental consequences."
>
> EPA's Scientific Advisory Board released a January 1986 study that echoed some of the same

concerns about data deficiencies and the present state of ecological knowledge. In addition, a growing list of ecologists, entomologists, population biologists [and evolutionary biologists,] continue to raise new concerns independently of official government bodies. (*EPA Journal* 12[1986]: 8–9)

A new set of ecological problems of a very different dimension and magnitude from those associated with petrochemical agripoisons will follow agribusiness's improper use of genetic engineering technology under uncontrolled environmental conditions. According to Dr. Martin Alexander, professor of agronomy at Cornell University and former chairman of the Recombinant DNA Study Group of the EPA Science Advisory Board:

Alien organisms that are inadvertently or deliberately introduced into natural environments may survive, they may grow, they may find a susceptible host or other environment, and they may do harm. I believe that the probability of all these events occurring is small, but I feel that it is likely that the consequences of this low-probability event may be enormous.

Agribusiness's technocratic attitude is one of absolute indifference to the environment, except as a resource, and it sees environmental protection as a constraint to economic growth. The basic laws of nature, of ecology, are ignored. Technocrats have such arrogance and such faith in their technology that they firmly believe that they can improve upon nature without understanding her. The natural laws of ecology are thus violated, even though it is well known that they are the basis of a sound and sustainable agriculture and economy. Industrial laws—economic paradigms of supply and demand and ever greater efficiency and productivity—replace natural laws.

But we have already learned from petrochemical-based agriculture that mainline, self-sustaining, "natural" agriculture is more profitable and healthful in the long term[15]. It is clear that without a total reevaluation of values and ethics, biotechnology will be misapplied and serious problems will arise.

It seems as though nature is always one step ahead of those who would control and exploit her but do not yet fully understand her. Perhaps with right understanding, right reverence, and action (as organic farming has always shown), we can live in harmony with nature and with one another. Perhaps then we will not need new bacterial pesticides to kill other living things and bacterial weapons to kill one another.

This needed shift in perception entails the realization that industrial principles cannot be applied to or imposed upon biological systems (such as fields, plants, and farm animals). These systems become dysfunctional and diseased, necessitating more research to correct the decline in productivity and efficiency. This focus—on treating the symptoms, not the cause—blinds us to the underlying first cause: our attitude toward and treatment of other living things, nature, and the environment.

Today's crops, livestock, and poultry suffer from a host of production-related diseases, reduced disease resistance, and metabolic disorders. Selective breeding and the unhealthy (pathogenic) environments in which crops and animals are raised make them susceptible to all kinds of pathogens—from bacteria and viruses to insects and worms. Genetic engineering will never fix these problems. At best, it will provide short-term profits by temporarily boosting efficiency and productivity, until these pathogens

[15] For example, see U.S. Department of Agriculture Study Team on Organic Farming, *Report and Recommendations on Organic Farming* (Washington, DC, 1980).

become resistant to genetically engineered bacterial pesticides and disease-resistant plants, animals, and animal vaccines, or until new pathogens arise.

Genetic engineers may be able to create corn plants and other crops that produce spider venom and other insecticides, and are resistant to attacks by certain insects, fungi, bacteria, and nematodes, but there is a high probability that such pests will quickly "learn" to deal with these obstacles, and new strains of pests will arise. Alternatively, entirely new species of pests might arise in response to ecological changes caused by a dramatic reduction in common pests. Though it may be claimed that making pest-resistant crops is a kind of "preventive medicine" that will decrease dependence on pesticides and herbicides, there is another point of view that genetic engineers have overlooked: It is possible that these disease problems arose as a result of ecologically unsound agricultural practices. Therefore, the right approach would be to change the practices rather than change the plants and make them adapt to conditions that are intrinsically unhealthy or pathogenic. There is a close analogy with the use of vaccines, antibiotics, and other drugs to make intensively raised farm animals resistant to diseases; the husbandry-related factors that cause animal stress, suffering, and increased susceptibility to disease are not addressed. The wise course would be to correct those environmental and husbandry factors that lead to increased susceptibility to disease in crops and livestock, rather than continue to treat the symptoms.

The inherent danger of genetic engineering is, as with other technologies, its *mi.* application. The potential breadth of its applications in agriculture is summarized in Table 4.1. Without a fundamental change in worldview from a technocratic to a more holistic ecological approach, the risks and costs of genetic engineering will far outweigh its benefits. And no government agency, or consortium thereof, will be able to regulate this industry effectively or predict and control its potentially adverse global impact on the environment or the structure of agriculture.

Human beings continue to play God. Each painful step is a revelation of ignorance. Genetic engineers make plants that can fix nitrogen in the soil themselves, eliminating the need for costly nitrogenous fertilizers. But what happens if the rate of fixation cannot be regulated by the plant? Nitrate overload could occur, increasing the plant's disease and pest susceptibility. Similar adverse consequences may occur in plants that are engineered to grow faster or produce more protein. Engineering plants that will grow in hot or cold climates to which they are not adapted may sound wonderful. But because these environments harbor insects and diseases with which the plants have not co-evolved to create a bio-stable relationship and ecology, more chemical poisons will be needed to protect them. Use of herbicide-resistant seeds will lead to more herbicide pollution of groundwaters.

Pathological instability will result from these short-sighted, profit-motivated (and in some instances altruistic, though misguided) innovations. Residual problems have stimulated the economy for some time now, but time, resources, and remedies are running out. The residual problems of an improperly applied and inadequately regulated biotechnology industry could become insurmountable.

If we continue to disturb the balance and forces of nature, the fertility, growth, and fecundity of crops and their primary and secondary consumers—wildlife, farm animals, and human beings—will be adversely affected. To try to rectify these problems through medical and agricultural bioengineering is to address the symptoms, not the underlying causes.

TABLE 4.1.

SOME POTENTIAL ENVIRONMENTAL APPLICATIONS OF GENETICALLY ENGINEERED ORGANISMS IN AGRICULTURE

Microorganisms

Bacteria as pesticides:

"Ice-minus" bacteria to reduce frost damage to agricultural crops.

Bacteria carrying *Bacillus thuringiensis* toxin to reduce loss of crops to dozens of insects.

Mycorrhizal fungi to increase plant growth rates by improving efficiency of root uptake of nutrients.

Nitrogen-fixing bacteria to increase nitrogen available to plants and decrease the need for fertilizers.

Viruses as pesticides:

Insect viruses with narrowed host specificity or increased virulence for use against specific agricultural insect pests, including cabbage looper, pine beauty moth, cutworms, and other pests.

Vaccines against animal diseases:

Swine pseudorabies
Swine rotavirus
Vesicular stomatitis (cattle)
Foot and mouth disease (cattle)
Bovine rotavirus
Rabies
Sheep foot rot
Infectious bronchitis virus (chickens)
Avian erythroblastosis
Sindbis virus (sheep, cattle, chickens)

Plants

Herbicide resistance or tolerance to:

Glyphostae
Atrazine
Imidazolinone
Bromoxynil
Phosphinotricin

Disease resistance to:
Crown gall disease (tobacco)
Tobacco mosaic virus

Pest resistance:
BT-toxin protected crops, including tobacco (principally as research tool) and tomato.
Seeds with enhanced antifeedant content to reduce losses to insects while in storage.

Enhanced tolerance to environmental factors, including:
Salt
Drought
Temperature
Heavy metals

Enhanced marine algae:
Algae enhanced to increase production of such compounds as B-carotene and agar or to enhance ability to sequester heavy metals (e.g., gold and cobalt) from seawater.

Forestry:
Trees engineered to be resistant to disease or herbicides, to grow faster, or to be more tolerant to environmental stresses.

Animals

Livestock and poultry:
Livestock species engineered to enhance weight gain or growth rates, reproductive performance, disease resistance, or coat characteristics.
Livestock animals engineered to function as producers for pharmaceutical drugs.

Fish:
Triploid salmon produced by heat shock for use as game fish in lakes and streams.
Fish with enhanced growth rates, cold tolerance, or disease resistance for use in aquaculture.
Triploid grass carp for use as aquatic weed control agents.

Source: Office of Technology Assessment, 1991.

Medical and Health Concerns

If we stand back and look at the appalling incidence of cancer, sterility, developmental defects, and inherited abnormalities in humans, we find a common thread. The answer lies not in the stars or in the genes, but in how our genes are affected by the contaminated food we eat, the water we drink, and the air we breathe. Tragically, some people are genetically more susceptible than others to agripoisons and industrial pollutants. Genetic engineering to correct these medical problems is a narrow (reductionistic) and instrumental (mechanistic) response to a problem that is fundamentally conceptual: namely, our attitude toward life and our mistreatment of the Earth, plants, and animals—and ourselves in the process.

It is a half-truth that improved perinatal and geriatric care contributes to the increased incidence of developmental disorders, cancer, and other degenerative diseases in humans by increasing survivability and longevity. The other half of the truth, which more and more "establishment" scientists and doctors are beginning to acknowledge, lies in the environment. Pets soaked in flea-killing pesticides have given birth to deformed offspring. Wild monkeys in a Japanese preserve that were fed regular human foodstuffs (because the preserve had an inadequate supply of natural food) have given birth to deformed offspring— herbicide residues in the food were the supposed cause. Genetic engineering is not the answer to similar health problems in humans, although it may well be applied profitably in the near future. We must cleanse the environment and correct our imbalanced and destructive relationship with it. As Albert Einstein foresaw in his uncompleted unification theory (and which the science of ecology has surely demonstrated), we do live in a unified field—self and environment are one, interpenetrating and interdependent. And it is highly destructive for us to act with total disregard of nature's unity and lawful harmony.

We need to recognize the tragic and ironic connection between biologically unsound agripractice and the medical industry. Agrichemicals, only 15 percent of which have been tested thoroughly for health risks, can cause genetic and developmental damage to fetuses and can cause cancer in laboratory animals, according to the National Academy of Sciences. These disorders are so common in people that the medical branch of genetic engineering is now planning to use biotechnology (and is investing lavishly in research) to correct genetic and developmental (somatic) diseases and cancer. This vicious circle of an unhealthy agriculture and an unhealthy population needs to be broken by a conceptual revolution in medical and agricultural practices.

But this is not to say that genetic engineering biotechnology has no role to play in human and veterinary medicine: Indeed it has, as in the development of safer and more effective drugs, vaccines, and diagnostics (see Table 4.2).

Genetically Engineered Foods and Food Security

The FDA will soon be granting permits for the sale of a variety of genetically engineered foods, notably genetically engineered fish (which grow faster with human, bovine, and salmon growth hormone genes) and dairy products from cows injected with bovine growth hormone (BGH). The Office of Technology Assessment has given carte blanche approval to BGH in terms of consumer safety and cow health.

TABLE 4.2.

BIOTECHNOLOGY PRODUCTS APPROVED BY FDA

DRUGS AND VACCINES

Product Name	Number	Indication(s)
Human insulin	1	Diabetes
Human growth hormone (HGH)	2	HGH deficiency in children
Interferon alpha*	3	Hairy cell leukemia; genital warts; AIDS-related Kaposi's sarcoma; non-A, non-B chronic hepatitis
Interferon gamma	1	Chronic granulomatous disease
Tissue plasminogen activator (TPA)	1	Acute myocardial infarction; acute pulmonary embolism
Erythropoietin (EPO)	1	Anemia of chronic renal failure patients; AZT-related anemia of HIV-infected patients
G-colony stimulating factor	1	Chemotherapy-induced neutropenia in nonmyeloid malignancy
GM-colony stimulating factor	1	Autologous bone marrow transplantation
Muromonab-CD3	1	Reversal of acute kidney transplant rejection
Hepatitis B vaccine	2	Hepatitis B

IN VITRO DIAGNOSTICS

Product Type	Number	Infectious Disease (Bacterial Viral, Fungal)	Tumor Marker	Analyte & Drug Monitoring	Blood Screening
Monoclonal antibody	571	127	2	433	9
DNA probe	53	42		11	
Recombinant DNA	13	11	1	1**	

FOOD INGREDIENTS		
Product Name	Number	Use
Chymosin	1	Cheese production

*Indications vary by company.
**Gene rearrangement.

The Gene Exchange newsletter recently published a menu of transgenic foods that we might soon find on our tables, especially if there are no labeling provisions that enable discriminating consumers to choose otherwise (see Figure 4.1).

Food industry biotechnologists—who are already selling irradiated and genetically engineered, laboratory-made analog foods and "nutrient supplements"—operate on the arrogant assumption that nutritional science knows all the answers and can quickly fix any problems that might arise. But they forget that human physiology and nutrition have evolved in concert over millennia. If we start changing these nutrient elements through biotechnology, it is likely to cause physiological and biochemical changes that may well be harmful to our health. Furthermore, the science of nutrition is still in its infancy; little is known about how various nutrients, antinutrients, and nonnutrients interact during the digestive process or about genetic, cultural, and age-related differences within the human population.

This is all the more reason for consumers to purchase natural foods and prepare their own meals from basic ingredients that have been neither processed, refined, nor genetically engineered. Analog foods are for analog humanoids, not for real people. Another reason for people to purchase real food is to keep farmers in business. It is not yet widely recognized that independent farmers are

Figure **4.1.**
TRANSGENIC FOODS MENU

---

*A DINNER OF TRANSGENIC FOODS**

APPETIZERS
Spiced Potatoes with Waxmoth genes
Juice of Tomatoes with Flounder gene

ENTREE
Blackened Catfish with Trout gene
Scalloped Potatoes with Chicken gene
Cornbread with Firefly gene

DESSERT
Rice Pudding with Pea gene

BEVERAGE
-Milk from Bovine Growth Hormone (BGH)-
Supplemented Cows

*Federal permits for environmental release are pending or
have been granted for all the transgenic plants and
animals included on the menu. BGH is under consideration
for approval as a veterinary drug.

Source: *The Gene Exchange*, December 1991.

becoming as endangered as natural foods. Profound structural changes are now taking place in agriculture and the related food industry as multinational corporations gain a monopolistic control over how and what farmers farm and what kinds of food people eat. Cary Fowler put it this way:

> What is now emerging throughout the corporate sector in the United States, Europe and Japan is a new, unprecedented institution of economic and political power; the multi-faceted transnational

life sciences conglomerate—a huge company that will use genes to fashion life-necessary products just as earlier corporate powers used land, minerals, or oil.[16]

The world's food system could become extremely vulnerable in many unforeseen ways if developments in agricultural biotechnology are not based upon the ethics and principles of humane, sustainable agriculture. Jack Doyle expresses this concern:

> Today, we may be moving toward a high-tech, house-of-cards agriculture worldwide, with genetic engineering at its base; a system in which one monkey wrench or one unforeseen mutation can create enormous problems. Just as the technology of hybrid corn production went wrong in 1970, aiding the advance of the corn blight, the agricultural biotechnologies of genes, microbes, and molecules might go wrong on a much grander scale in the future. Despite what its proponents may claim for it, this is not an invincible or fail-safe technology.[17]

The reality of a rising, market-driven, global agricultural biotechnocracy must be faced today if the democratic principles of a socially just, humane, and sustainable post-industrial society are to prevail. The ethical, environmental, social, and economic concerns of new developments in biotechnology must be addressed. They cannot be left to government bureaucrats and multinational corporate representatives, whose narrow vision of economic growth

---

[16]   C. Fowler, "The Laws of the Land: Another Development and the New Biotechnologies." *Development Dialogue* I(2): 49(1988).

[17]   J. Doyle, *Altered Harvest* (New York: Viking Press, 1985).

as the ultimate social good precludes such considerations, except insofar as they may interfere with the "national interest" and corporate profits (which underlie their interpretation of sustainability). For these democratic principles to prevail in the face of an antidemocratic technocracy, there must be a strong "grass roots" alliance between consumers and those farmers, food wholesalers, and retailers who endorse these principles and put them into practice.

By aligning itself with sustainable agricultural practices, biotechnology would be on a far less vulnerable foundation than if it were integrated with conventional, ecologically unsound agriculture. A cardinal feature of sustainable agriculture and food security is genetic diversity, in terms of both the kinds and varieties of crops and livestock being raised and how various crops and farm animals are husbanded. For example, the conventional, corn-fed, high-yielding Holstein/Friesian dairy cow is now genetically less suited for a high-forage-based diet. Dairy farmers who would like to switch to ecologically superior forage-crop rotations can't find good cows whose ancestors did well on grass, because selective breeding has focused on developing dairy cows that won't give much milk if they don't have a lot of corn and other "concentrates" in their rations. Likewise, the selective breeding of lean pigs with hardly any back fat now means that hog farmers will be hard-pressed to find breeding stock better suited for outdoor living (with more fat insulation on their bodies) if they want to switch to more natural, crop-integrated outdoor hog production.

Industrial agriculture focuses on monocrop or mono-culture farming—vast fields of the same crop and long sheds full of animals—and, as we have learned, these systems are dysfunctional. One cannot impose an industrial paradigm on complex biological systems without a deep understanding and respect for those systems.

International Concerns

There are many humanitarian concerns that are generally overlooked by the media, the public, and many involved in the biotechnology industry. Developing new artificial sweeteners, flavors, and high-yield crops, for example, may seem laudable. But the technocratic ideologies of increased productivity, efficiency, monopoly, and profitability can have harmful long-term consequences to other world-market sectors. To lapse into the neo-Darwinian rationalization that market competition is natural and "healthy"—insofar as the best and fittest survive and such competition stimulates innovation and productivity—is to be ethically blind. A technocracy—more specifically a biotechnocracy—that is blind to humanitarian issues, especially the long-term adverse social and economic consequences of new applications of biotechnology, is cause for concern. (See Table 4.3 for a list of biotechnology applications proposed to solve the third world's livestock production problems.)

The Rural Advancement Fund International (RAFI)[18] has evaluated the potentially adverse consequences of biotechnology on developing nations. Concerning the use of genetic engineering to make plants tolerant of the damaging effects of herbicides, RAFI presents the following summary (*RAFI Communique*, November 1987):

> **IMPACT**: Industry focus on the development of herbicide tolerant crops indicates that instead of ending the chemical era in agriculture, biotechnology will be used to extend it. Herbicide tolerance could lead to an increase in the farmers' cost of production; greater risk for agricultural

---

[18]   P.O. Box 665, Pittsboro, NC 27312.

TABLE 4.3.

POSSIBLE APPLICATIONS OF BIOTECHNOLOGY TO THE SOLUTION OF LIVESTOCK PRODUCTION PROBLEMS IN THE THIRD WORLD

Problem	Possible Biotechnology Solution	Scale of Economic Impact	Probable Time to Commercial Use*
Animal, poultry, fish diseases	New vaccines	Large	Short
	New diagnostics	Moderate	Short
Poor quality of forages	Microbial treatment of forages	Moderate	Medium
	Modification of rumen microflora	Moderate/ large	Long
	Genetic improvement of forages and their symbionts	Moderate	Medium
Difficulty of implementing selection programs	Selection in nucleus herds, using ET, sexing	Large	Medium
	Use of RFLP markers to assist selection	Moderate	Medium
Difficulty of maintaining dairy cattle performance after F1 cross	Use of IVF, ET, and sexing	Large	Long
Cost and environmental challenge to imported cattle	Use of ET to import embyros	Small	Short
Need for increased efficiency in intensive systems	Use of rBST and rPST in dairy and pig production	Large	Short

Sources: Cunningham (1990); Doyle and Spradbrow (1990).
ET: embryo transfer.
IVF: in vitro fertilization.
rBST, rPST: recombinant bovine or porcine somatotropin (growth hormones produced using recombinant DNA technology).
RFLP: restriction fragment length polymorphism, or direct DNA typing of individuals.
*Short: now or before 5 years; medium: 5 to 10 years; long: over 10 years.

workers; increased environmental damage (especially groundwater contamination); more chemical residues in the food chain; danger of crop loss.
WHEN: Early 1990s.
COUNTRIES AFFECTED: All countries.
PARTICIPANTS: At least 28 enterprises have launched over 65 research programs focusing on herbicide tolerant crop varieties. These include major agrichemical companies: Monsanto, DuPont, Ciba-Geigy, ICI, Rhone-Poulenc, Bayer, Hoechst, and more.
ECONOMIC STAKES: Market value is expected to exceed $3.1 billion by the mid-nineties and touch $6 billion by the turn of the century.

In a subsequent *Communique* (December 1987) dealing with the impact of the development of artificial seeds (by so-called somatic embryogenesis to make seeds capable of producing their own fertilizer and pesticides), RAFI's concerns were as follows:

IMPACT: The integration of all agricultural inputs could lead to an increase in the farmers' cost of production; jeopardize the health of agricultural workers; and increase environmental damage and chemical residues in the food chain.
COUNTRIES AFFECTED: All countries, but the economic impact may first be felt in Chile, Mexico, Morocco, New Zealand, Tunisia, and Tanzania (major areas for seed multiplication).
WHEN: Early 1990s.
PARTICIPANTS: Monsanto, DuPont, Ciba-Geigy, ICI, Rohm & Haas, Rhone Poulenc, American Cyanamid, Hoechst, and other leading chemical companies.
ECONOMIC STAKES: Integration of the $50 billion seed, pesticides, and fertilizer industries into the genetics supply industry. Increased market value may equal $12.1 billion by the year 2000.

RAFI is also evaluating the adverse consequences of other new developments in biotechnology on the economies of developing countries. The following summary statements from periodic *Communiques* show the serious social and economic consequences these developments could have:

New Substitutes Threaten to Displace Export Market for Water-Soluble Gums (September 1986)

ISSUE: Replacement of major cash crop.
COUNTRIES AFFECTED: Nigeria,   Senegal, Sudan, and others.
CROP: Gum arabic ("hashab").
IMPACT: Possible loss of $60 million in annual export earnings and seasonal employment.
WHEN: Immediate; 1986 crop.

Vanilla and Biotechnology (January 1987)

ISSUE: Natural vanilla production via tissue culture technology.
CROP: *Vanilla planifolia*—the commercially important species of vanilla orchids.
COUNTRIES AFFECTED: Madagascar, Comoros Islands, Reunion, Indonesia.
IMPACT: Possible loss of up to $67 million in annual export earnings.
WHEN: Mid-1989.
COMPANIES INVOLVED: David Michaels Co., Inc.; Escagen Corp. (formerly International Plant Research Institute).

Biotechnology and Natural Sweeteners— Thaumatin (February 1987)

ISSUE: The use of biotechnology to produce the intensely sweet thaumatin protein.
PLANT: Thaumatin is derived from the fruit of a West African rain forest shrub.

COUNTRIES AFFECTED: Product will be marketed as a low-calorie sweetener in Europe, Japan, and the U.S.

IMPACT: In combination with other newly developed sweeteners, these products offer the potential to erode traditional sugar markets.

WHEN: A genetically engineered thaumatin sweetener is now being produced in the laboratory; one company will apply for U.S. regulatory approval in 1988–89.

COMPANIES INVOLVED: Unilever (the Netherlands); INGENE for Beatrice Foods (USA); (unconfirmed: DNA Plant Technology, Inc. for Monsanto, USA).

Cacao and Biotechnology:
A Report on Work in Progress (May 1987)

ISSUE: Cacao and biotechnology.

CROP: *Theobroma cacao.*

COUNTRIES AFFECTED: All cacao-producing countries of the third world—especially Ivory Coast, Ghana, Brazil, Cameroon, Nigeria, Malaysia, and Ecuador.

IMPACT: Development of high-yielding cacao varieties could lead to overproduction and jeopardize price and stability of cacao-producing countries while shifting production from small-scale producers to large-scale plantations; the use of biotechnology to convert low-priced oils into cacao butter could drastically reduce the demand and price for cacao beans.

COMPANIES INVOLVED: U.S. Chocolate Manufacturer's Association (15 U.S.-based companies) and the American Cocoa Research Institute; Hershey Foods; DNA Plant Technology; Genencor; CPC International; Ajinomoto (Japan); Fuji Oil (Japan); Cadbury-Schweppes (United Kindom).

WHEN: Work on all areas is now in progress.

The consequences of biotechnology extend far beyond society and agriculture; the entire animal kingdom is also affected—both directly and indirectly. The next chapter documents what genetic engineers are doing directly to animals, purportedly for the benefit of society and agriculture, and the adverse impact on animals, wild and tame.

CHAPTER 5

GENETIC
ENGINEERING
OF ANIMALS

A cow is nothing but cells on the hoof.
—THOMAS WAGNER, veterinarian
biotechnologist (*Fortune Magazine*, October
1987, p. 80)

*Q*uite apart from the ethics and wisdom of trans-
forming the natural world into a human world,
which our power over the gene now enables us to do, is the
morality of turning animals into biological machines. It
may be morally and ethically acceptable to turn bacteria
into machines for the manufacture of various hormones
and other biological chemicals and to enhance the utility
of various plant species (negative environmental consequences notwithstanding), because these living things

are not sentient. They lack the capacity to suffer, to experience pain and emotional distress. If they were sentient, suffering could well result from the effects of various genetic manipulations on their body structure and physiology. So is it morally and ethically acceptable to turn animals such as mice, pigs, and sheep, which are sentient, into biomachines for the manufacture of protein and other biological materials? Before the perfection of gene-insertion and -deletion techniques and the development of the desired animal machines, there have been accidents, deformed and defective creatures born, their psyches imprisoned in alien bodies. Already giant mice have been created by inserting the growth-regulating genes of rats and humans into them while they are embryos. Should we care how a mouse, made giant, might feel? I believe that we should, for this is the very quality that makes us human.

Transgenic Animals

This is just the beginning. The U.S. Department of Agriculture (USDA) has used public funds to apply this same technique of inserting human genes into animals to try to create giant pigs and sheep. Of the USDA's "superpigs" that carry the human growth gene, only one in every 200 embryos survived. And those pigs that lived to maturity had impaired vision, were arthritic and lethargic, and were prone to pneumonia because their immune systems were dysfunctional.

Embryogen Biotech Company in Athens, Ohio, has succeeded in inserting the growth-regulating gene of cattle into pig embryos, with the hope of creating a leaner and larger variety of pig (*Los Angeles Times*, 12 July 1987). Pigs with little or no protective body fat would have to be raised indoors in confinement "factories," where humane conditions to date are virtually nonexistent. Animals genetically designed to live indoors will, of course, mean the perpetuation of capital-intensive factory farms.

Genetically engineering livestock designed to tolerate extreme climatic conditions, from the tropics of equatorial countries to the cold tundra of the polar and high-altitude regions of the world, could mean the extinction of existing wildlife species in these areas. They would be displaced and exterminated by competing livestock interests, as is happening in Amazonia, Africa, Australia, and Asia today.

According to *New Scientist* magazine (28 April 1988, p. 27), biotechnologists at the University of Adelaide in Australia have transgenic pigs in their seventh generation. They carry an extra growth hormone gene that makes them 30 percent more efficient and brings them to market seven weeks earlier than normal pigs. Plans are under way to patent these animals, along with transgenic sheep developed by the Australian Commonwealth Scientific and Industrial Research Organization that purportedly grow 30 percent faster than normal sheep. This organization is also preparing to transplant two genes into sheep that it hopes will speed the growth of wool. These genes will cause the sheep's liver to produce enzymes that convert sulfur-bearing compounds (normally lost as gas in the intestinal tract) into the amino acid methionine, which increases wool growth. Australian scientists are also trying to "perfect" a genetically engineered hormone that will make sheep shed their fleece, cutting the costs of shearing. Tests to date on pregnant sheep show that the hormone causes some of them to abort.

Ohio State researcher J. Mintz (who has successfully inserted rabbit growth genes into mouse embryos to create mice that grew two and a half times larger than normal) has predicted the development of cattle weighing over 10,000 pounds and pigs that are twelve feet long and five feet high. Such monstrosities of utility are within the realm of possibility within the next ten to twenty years, though he cautions that such genetic engineering innovations might not be desirable for economic, environmental, anatomical, institutional, and ethical reasons.[1]

Scientists in the United States, Japan, Europe, and Australia have created a number of transgenic animals— pigs, lambs, calves, and fish—containing the genes of other species, such as the human and bovine growth hormone genes. Success rates of gene insertion are extremely low, and the entire process is time-consuming and costly. Many of the funds come from the public by means of government tax revenues.

Some researchers have recently opted to put extra growth-regulating genes of sheep origin rather than human origin into lambs because they felt that this would be more acceptable to laypersons, especially consumers. However, even though these lambs were leaner, they did not have increased feed efficiency, they were diabetic, and they had such severe health problems that they died before reaching puberty. The cause of death varied, but it was clear that the growth hormone can adversely affect the liver, kidneys, and heart.

Australian government scientists are planning to engineer sheep to secrete insect repellent from their hair follicles. Their goals are to keep sheep free of fleece-damaging blowflies and to produce the world's first mothproof wool.

Merck & Co., the European-based pharmaceutical company, has applied for a patent in the United Kingdom on its superchicken, called Macro-Chicken. Merck has developed a line of broiler chickens that carry the growth

[1]   "Biotechnology," *Venture Magazine*, February 1984.

gene of cows. It hopes to corner the market with a highly feed-efficient, fast-growing bird.

It is likely that Merck's Macro-Chickens will have a variety of health problems too. But if the birds eat well and grow quickly, they will be ready for slaughter before severe health problems ever develop. But what of the reserve breeding stock of transgenic chickens that will not be raised for slaughter—will they suffer? Because such information is proprietary, corporations are not likely to reveal the limitations and risks of their new patentable creations. Trade secrets notwithstanding, the social and economic consequences of creating transgenic farm animals have been given scant attention.

Critics of the genetic engineering of farm animals question the use of public funds to make these animals produce more meat (even if it *is* lean). The short- and long-term costs of such research are not considered. A major problem of contemporary intensive animal agriculture is overproduction; meat and milk surpluses are already a chronic problem in some industrialized countries. It is unlikely that the creation of transgenic farm animals will help feed the hungry world, since meat production efficiency has built-in limitations and inevitable environmental costs.

Genetic engineering technology is being used in an attempt to alter sheep's and cow's milk so that it can be consumed by a large percentage of the world population that is lactose intolerant. This may be a more fruitful approach to feeding the hungry, since milk production is far more efficient, ecologically sound, and cost-effective than meat production, with or without biotechnology.

Human genes responsible for the production of proteins in mother's milk are being inserted into calf embryos with the hope of creating a new generation of cows that produce "humanized" milk.

Molecular Farming

Research in the field of molecular or pharmaceutical farming—called "pharming"—is progressing rapidly, which means that this may soon become a major new industry. For example, Dr. A. J. Clark and associates have successfully injected mouse embryos with a gene from sheep that makes the mammary glands manufacture beta-lactoglobulin, a substance that mice do not normally produce in their milk but sheep do.[2] This research team has also succeeded in genetically engineering sheep by inserting a human gene into sheep embryos, causing them to produce Factor IX (a blood component that hemophiliacs lack and must receive in medical treatment) in their milk. Similar research on farm animals is being conducted in the United States at the School of Veterinary Medicine, University of Pennsylvania, where attempts are being made to implant the human metallothionein growth hormone gene and the human blood clotting Factor IX gene in a variety of animals, including rabbits, sheep, goats, pigs, and cows. Researchers at Genzyme and Tufts University have developed the first genetically engineered mice that produce a membrane protein for possible human pharmaceutical use in their milk. The discovery provides a means to produce large quantities of the membrane protein cystic fibrosis transmembrane regulator (CFTR), the protein coded by the gene associated with the genetic disease cystic fibrosis, as well as other therapeutic proteins. Biotechnologists at Virginia Polytechnic and State University have created transgenic pigs that produce in their milk protein C, an anticoagulant that may one day help heart attack and stroke victims.

None of these new animal creations have yet provided any medical benefits to humans, but venture

---

[2] *The Washington Post*, 10 August 1987; *Nature*, 6 August 1987.

capitalists are investing in this line of research and development. Recently a biotechnology company—DNX Inc., of Princeton, New Jersey—reported that it has developed a line of transgenic pigs that produce human hemoglobin. But we are still a long way from having farmers raise pigs to be human blood donors.

It should be emphasized that most genetic engineering research in farm animals has focused on increasing productivity. Research on increasing resistance to disease through genetic engineering is still in its infancy. Such research should be questioned, since improvements in farm animal husbandry are tried and true and are surely more cost-effective ways of improving animal health and well-being. U.S. government scientists are now embarking on mapping the genetic makeup of cattle and pigs, but the relevance of this long-term and costly project to sustainable agriculture is unclear, since the primary focus will be to enhance productivity traits in concert with current intensive systems of livestock production.

Other Innovations

Other developments in farm animal biotechnology (which do not entail gene transfer) can have profound social and economic ramifications. These include the development of cow clones and a technique to preselect the sex of offspring. Still, biotechnologists cannot explain why many cloned calves are almost twice normal size at birth and must be delivered by Caesarian section.

Although no plant genes have been inserted into animals, animal genes have been successfully incorporated into the genetic structure of various plants. Tobacco plants have been successfully implanted to produce functional human antibodies that can be used for diagnosing and treating human diseases. The "antifreeze" gene of the flounder, which produces a protein to stop the fish from freezing, has been cloned and inserted into tomatoes and tobacco. In the future, crops may be protected from frost by fish genes.

Pharmaceutical "pharming,"using transgenic plants, is probably the most ethically (if not also ecologically) acceptable of all the new developments in agriculture. It is a humane and less costly alternative to using transgenic farm animals for such ends. (But it will mean more competition with the livestock and poultry industries for arable land, so much of which is squandered today to grow animal feed.) Tobacco plants have recently been bio-engineered to produce a food-processing enzyme, amylase, which converts starches to sugars, and to manufacture Compound Q (tricosanthin), an antiviral compound used to treat AIDS patients. Such products from whole plant (and animal) "bioreactors" are claimed to be superior to recombinant bacterial systems because these bioreactor systems may be cheaper than cell fermentation and produce proteins in the more desirable glycosylated form. Other potential crops for the molecular "pharmers" of the future reviewed in *Genetic Engineering News* (December 1991, p. 54) include high-value enzymes used in food processing, medical proteins and peptides like the alpha and beta chains of hemoglobin, human insulin, and serum albumin.

Since fish farming is on the increase, biotechnologists have been busy developing "superfish" by inserting growth hormone genes from humans, cattle, chickens, mice, and other fish into a variety of commercially raised fish such as carp, rainbow trout, catfish, Atlantic salmon, walleye, and northern pike. The antifreeze gene of the winter flounder is also being inserted into other fish species to expand commercial fish production in cold regions and seasons.

A biotechnologist at the Army Research Laboratory in Natick, Massachusetts, has cloned the silk-producing gene of the golden orb weaver spider and spliced it into bacteria, which in turn produce large quantities of spider silk protein. Stronger than silkworm silk and even steel, this new product may have wide commercial use, especially to develop new fabric for bulletproof vests, helmets, parachute cords, and other equipment that must be both strong and light.

On the brave-new-world frontier of medicine, scientists have created a variety of transgenic mice. A line of mice has been created that carries human genes that result in deformed red blood cells, providing a new model for sickle-cell anemia. Rats have been developed that carry the defective human gene HLA-B27, which causes a painfully crippling form of arthritis. The clinical relevance of these new creations has yet to be demonstrated. There is no foreseeable benefit to animals of making them transgenic, except perhaps for endangered and genetically "fragile" or defective species such as the cheetah and South American maned wolf—species whose survival may depend on genetic alteration.

Another development is a gunpowder-charged 'bio-blaster' that literally shoots genetic material into microbes, plants, and animals—a crude technique funded in part by "Du Pont/ConAgra Vision." (*Bio/Technology* 10:1992 286–291).

Although the debate over using primates such as chimpanzees in biomedical research as models for various human diseases continues, the more economical, if not more ethically acceptable, alternative of using transgenic mice instead of primates is gaining considerable acceptance. Furthermore, transgenic mice hold the potential for reducing the number of animals needed for various studies, such as screening harmful chemicals, because they are especially sensitive and are more specific models than most regular varieties of mice used in biomedical research.

Research is continuing on the identification of genes responsible for various inherited diseases, especially in purebred dogs and livestock, and on genes that play a role in development, growth, milk and egg production, disease resistance, and other physiological processes. The results of such costly research may eventually be of benefit to animals in terms of their health and overall well-being. But the benefits will be limited if this approach becomes overly reductionistic and utilitarian and is not integrated with a more holistic, if not traditional, approach to improving animal health and well-being. This is especially true if the focus of such research is primarily on enhancing the exploitative value of animals.

The human genome is being sequenced and genetic defects and strengths identified. Next will be the cow, pig, and dog. All to what end? New medical and veterinary products and services will certainly result. But genetic determinism can lead to eugenics, the science of making "perfect" offspring. And eugenics means genetic imperialism. Do we really want or need a creation made over into a human image of perfect utility? As Dr. Lester Ichinose, scientific advisor to the National Antivivisection Society, sees it, "*Homo sapiens* is evolving into the ultimate parasite, transgenically infecting the life around it not only with its biology, but also with its hopes and fears."

Animal Welfare

One potential benefit of biotechnology to animals is in the development of genetically engineered vaccines, hormones, immune-system enhancers, birth-control regulators, and diagnostic and screening tests. However, this new generation of veterinary products and services may be a mixed blessing. It is not without potentially adverse animal health, socioeconomic, and ecological consequences —as with the bovine growth hormone. Many of these products are no substitute for humane animal husbandry, sound breeding, and good nutrition.

In January 1992, the European Parliament agreed to extend a moratorium on the bovine growth hormone BST, pending further evaluation of its safety in terms of animal welfare. Concerns were also expressed over a potential consumer boycott that would harm the dairy industry, and over the "fourth hurdle", social and economic implications, assessment of new products on economic and social grounds. This moratorium was triggered in part by evidence of a cover-up of Monsanto-sponsored research on BST in dairy cows at the University of Vermont. These cows apparently gave birth to stillborn and deformed calves, often retained the placenta after giving birth, and suffered more frequently from ketosis, a metabolic disease symptomatic of excess breakdown of fat. Veterinarian David Kronfeld believes that BST may actually cause genetic damage to the offspring of treated calves.[3]

Because of prohibitive cost limitations, genetic engineering of animals presents a potentially serious animal-welfare concern. In today's factory farms it is too costly to design systems that would fully provide for the animals' overall welfare. So too will it be too costly to correctively redesign animals physically and psychologically to better adapt them to any harmful consequences of genetic engineering. Further genetic engineering to improve animals' welfare will not be cost-effective, since the primary goals of livestock and poultry genetic engineering are production and profit oriented. We have learned this sad fact from

---

[3]   D. MacKenzie, "Doubts Over Animal Health Delay Milk Hormone." *New Scientist*, 18 January 1992, p. 13.

factory farming, where animals' welfare, and their basic freedom to develop, express and experience their intrinsic nature, are frustrated, truncated, and impaired.

Proponents of genetic engineering argue that there are no fundamental differences between these new techniques of genetic engineering and the old method of selective breeding. This rationalization ignores the fact that there are genetic barriers between animal species that prevent interbreeding and the exchange of genes from one species to another (transgenic engineering), probably for good reason. This is one of nature's laws that may be imprudent to ignore. Furthermore, traditional selective breeding of farm animals to enhance egg and milk production and growth has contributed to widespread suffering, increased susceptibility to infection, and new, complex diseases in factory-farmed animals. These so-called production diseases, which are well recognized by animal scientists and veterinarians, have been documented in my book *Farm Animals: Husbandry, Behavior and Veterinary Practice* (Baltimore: University Park Press, 1984). In order to offset financial losses from these production-related diseases and the stress and suffering to which farm animals are subjected in overcrowded "superfarm" factories, antibiotics and other drugs are needed. This is now recognized as a serious hazard to consumer health.

Given that genetic manipulation of farm animals by natural means (selective breeding) to enhance productivity and efficiency has resulted in widespread animal suffering and sickness primarily for reasons of expedience and profit, it is unlikely that genetic engineering of farm animals for the same reasons will contribute to their health or well-being. Today their health and well-being are sacrificed for overall productive efficiency and profitability, and tomorrow will be no different. Nonrenewable resources (topsoil, water, and fossil fuels) will become even scarcer and more costly, the price of animal feedstuffs will increase, and farmers will experience even greater economic pressures, which will force them to further sacrifice animals' health and well-being in order to turn a profit. Ironically, industrial agriculture is already too successful in that it is overproductive. Surpluses of meat, eggs, and dairy products are a chronic problem in the United States and other developed nations. Those future farmers that have "super" animals—animals that grow twice as big twice as fast, or produce more milk or eggs or offspring—will have an economic edge over farmers who do not have such stock. Another competitive economic treadmill will thus arise, and a new market will be created for these animals, along with another mountain of surplus products.

There is currently some interest in putting genetically engineered bacteria into the disgestive systems of farm animals that will break down material the animals could not otherwise digest and convert it into meat, eggs, or milk. But such changes in the internal ecology of the animals' digestive systems increase the probability of new disease problems and more animal suffering.

Animal-rights philosophy holds that animals have inherent value, needs, and interests quite independent of their value and usefulness to us. If we exploit animals to satisfy our own needs (even if we have "created," bred, and raised these animals ourselves), then we should give them equal and fair consideration. It is morally wrong to violate an animal's right and entitlement to humane treatment. This ethic is written into law under the Federal Animal Welfare Act and state anticruelty statutes. Since the genetic engineering of animals may cause them to suffer from structural and physiological changes that have been deliberately, accidentally, or coincidentally induced by genetic manipulation, it is surely unethical and a violation of humane ethics and legal statutes to subject animals to such manipulation.

Introducing the genes of one species into another, regardless of potential animal suffering, also raises the ethical issues of violating the sanctity and dignity of the life of the individual animal and endangering the integrity and continuation of its species. Sheep at the British agricultural research station in Cambridge and at the University of California at Davis have been given the heads of goats by means of embryonic microsurgery. Goats' minds attached to sheep's bodies—for what purpose?

Many people are horrified by such demonstrations of scientific serendipity: Giant mice and goat-headed sheep are just the beginnings of a new age of biotechnology where human dominion over the rest of creation will be absolute. Is it not hubris—and biological fascism—to regard and treat animals and other living things as though they have been created primarily for our own exclusive use?

Some people claim with religious conviction that this is not hubris but God's will. They cite Genesis as proof that we have been given dominion over the rest of creation. The original meaning of the word *dominion* comes from the Hebrew root verb *yorade*, "to go down," which implies sympathy and communion—humane stewardship—rather than exploitative domination. When we recognize our commonality with the animal kingdom, we will be able to steward with compassion and have regard for the life of the beast. This is the philosophical basis of today's animal-protection and "deep" ecology movements.

Animal Suffering

A major concern of all humanitarians is whether genetic engineering will cause animals to suffer. The answer is being evaded by proponents of biotechnology who claim that "unnecessary" suffering will be avoided and that existing federal animal-care guidelines and regulations will take care of the problem. But transgenic engineering of animals has already caused them to suffer, even though this was not anticipated by the researchers. Federal animal-welfare regulations contain no reference to genetically engineered animals and cover only the care of animals. They say nothing about preventing or alleviating animal suffering following genetic reprogramming.

Animal suffering has already resulted from genetic engineering, and as more animals are subjected to transgenic intervention (or reprogramming) the probability of animal suffering will increase. There are several different types and sources of suffering, which are detailed below.

DEVELOPMENTAL ABNORMALITIES.   Following gene insertion into embryos, the embryos often fail to develop normally. Some may die in utero and be aborted or resorbed, or some may be born with a variety of developmental defects, sometimes due to what is termed *insertional mutations.* Because these defects may not be manifested until later in life, there can be no accurate prediction of whether engineered animals will suffer. And because of the nature of genetic reprogramming, there can be no safeguards to prevent animal suffering in the first place. These problems are to be expected in the initial phase of creating transgenic animals and in other genetic manipulations. The new technique called homologous recombination, where a mutated gene can be inserted into the correct place on a chromosome, and another relatively new method of cultivating embryonic stem cells and reinserting them into the mouse embryo, have greatly reduced the inaccuracies of earlier "hit or miss" gene insertion by microinjection.

Several developmental abnormalities and diseases, such as cancer, have been caused deliberately by using technology to create animal "models" of various human disorders. Strict guidelines are needed to minimize animal suffering, especially since the most commonly used species —the mouse—is excluded from protection under the Federal Animal Welfare Act.

W. French Anderson has emphasized that the micro-injection of eggs with foreign genes "can produce deleterious results because there is no control over where the injected DNA will reintegrate in the genome."[4] This can mean that a gene for a certain hormone or other protein may express itself in an inappropriate tissue. He notes that "there have been several cases reported where integration of microinjected DNA has resulted in a pathological condition." This is one of the major reasons that Anderson is opposed to transgenic germ-line therapy in humans. And it is a valid reason for concern over the welfare of animals subjected to this kind of treatment in early embryonic life.

Once the anticipated genetic changes have been accomplished and the new animal prototypes developed as foundation breeding stock, additional problems can be anticipated. These have already been shown to occur in transgenic animals.

DELETERIOUS PLEIOTROPIC EFFECTS. This term refers to multiple harmful effects by one or more genes on an animal's phenotype. The phenotype is the entire physical, biochemical, and physiological makeup of an individual. These problems are now well recognized by biotechnologists.

The well-publicized health problems of the USDA's transgenic pigs that carry the human growth gene were unexpected, since mice and rabbits reprogrammed with

---

[4]   *The Journal of Medicine and Philosophy* 10(1985): 275–291.

this same gene did not manifest deleterious pleiotropic effects to anywhere near the same degree.

These pigs were arthritic, lethargic, and had defective vision arising from abnormal skull growth. In addition, they did not grow twice as big twice as fast, which was the expected result based on the effects of the human growth gene in mice. These pigs had high mortality rates and were especially prone to gastric ulcers and pneumonia. The conclusion was that genetic change had seriously impaired their immune systems.[5]

This illustrates that pervasive suffering can arise from genetic engineering. It also demonstrates another principle: The fact that a genetic change in one species causes little apparent sickness and suffering does not mean that the same genetic change in another species will have comparable consequences. *Predictions and assurances about the safety and humaneness of genetic engineering cannot be generalized from one animal species to another.*

NEW HEALTH PROBLEMS. As the biotechnologists themselves have shown, new health problems following genetic reprogramming can arise, resulting in animal sickness and death. For example, pigs treated with somatotropin to stimulate growth require additional dietary lysine and possibly other essential amino acids according to Kansas State University researchers (*Feedstuffs*, 7 December 1987). What this research indicates is that special diets and other health-corrective treatments will be needed following some forms of genetic reprogramming. And veterinary medical knowledge will be inadequate to deal with the special requirements of animals subjected to genetic reprogramming. Additional research will

---

[5]   Caird Rexroad of the USDA's Beltsville Research Center reported that transgenic sheep carrying human growth hormones "did well for 180 days and then became unhealthy."

be needed to correct health problems and associated suffering.

DISEASE RESISTANCE. Biotechnologists contend that through genetic engineering animals can be made disease resistant, which will help reduce animal suffering (shipping fever in cattle, which causes considerable economic loss to the livestock industry, is a commonly used example). However, the notion that genetically engineered disease resistance will reduce animal suffering is scientifically naive because it reflects a single-cause (bacteria/virus) approach to disease. Simply endowing an animal with resistance to a particular disease will not protect it from other pathogens or from the stress factors and contingent suffering that make it susceptible to disease in the first place, such as transportation stress and overcrowding.

ERRONEOUS "BENEFITS" TO ANIMALS. There are other erroneous claims of the potential benefits of biotechnology to the animals themselves. It has been claimed that genetic engineering could be used to help cure animals of genetic disorders (almost 400 diseases of genetic origin have been identified in highly inbred purebred dogs, and there are dozens that afflict other domesticated species). But simply stopping the practice of inbreeding, and not breeding defective animals, is a better solution.

It has been claimed that genetically engineering livestock to be resistant to various tropical diseases (such as sleeping sickness) and to extremes in climate would benefit them as well as the industry. But it would not benefit other animals (i.e., threatened and endangered wildlife species) that are displaced and exterminated because their habitats are taken over by the livestock industry.

PRODUCTIVITY. Using genetic engineering biotechnology to increase the productivity and efficiency of farm animals (e.g., growth rates, milk or egg yield) *will increase the severity and incidence of animal suffering and sickness*. It is already extensively documented that farm animals raised under intensive confinement husbandry systems in order to maximize production and efficiency suffer from a variety of production-related diseases. By using the term *production-related diseases*, animal scientists acknowledge that animal sickness and suffering are an unavoidable and integral aspect of modern livestock and poultry farming. Using biotechnology to make animals even more productive and efficient under these conditions will place their overall welfare in greater jeopardy than ever, because the severity and incidence of production-related diseases will increase.

SELECTIVE BREEDING. Biotechnologists argue that genetic engineering is simply an extension of selective breeding and that since mutations (spontaneous genetic changes) occur naturally, there is nothing morally wrong or unethical about altering animals through genetic engineering. However, they totally ignore the scientific and medical evidence of the harmful consequences of deliberate genetic manipulation, as evidenced by the wide variety of disorders in purebred dogs.

As previously pointed out, the domesticated dog, our closest and oldest animal companion, is afflicted by almost 400 diseases of hereditary origin. These have been produced (like the genetic disorders of farm animals) through traditional selective breeding and inbreeding procedures in order to "fix" various traits for reasons of utility and esthetics. Without the careful attention of their owners and veterinary expertise, many dogs afflicted with these diseases would never survive to sexual maturity or be able to breed or raise their offspring successfully.

So-called semilethal inherited traits are those that can cause dogs and other domesticated animals considerable sickness and suffering or chronic discomfort. Under natural (i.e., wild) conditions, animals so afflicted would soon die due to the rigors of natural selection. Some of these traits in dogs and other domesticated animals are actual *mutations*. Examples in the dog include the deformed face of the Pekingese and bulldog, the pendulous ears of the cocker spaniel, the skin folds and wrinkles of the Shar-pei, giantism in the Irish wolfhound and Great Dane, and achondroplastic dwarfism in the basset hound and dachshund. Other lethal and semilethal inherited diseases common in many breeds are expressed structurally or physiologically, such as hip dysplasia, glaucoma, epilepsy, hemophilia, and immune-system dysfunction. Some diseases are linked with selectively bred-for mutations, such as the merle coat color of the Shetland sheepdog and albinism in the bull terrier, which are associated with blindness and deafness, respectively.

Veterinary medical research has only recently begun to recognize that so many of the health problems of companion animals are genetic in origin. Few of the disorders that affect farm animals have been looked at from this perspective. But the research that has been done reveals the same trend: an increasing incidence of health problems, many of them genetic in origin, arising as a consequence of selective breeding for reasons of utility. The genetic engineering of farm animals for these same purposes will therefore have similar consequences—an increased incidence of structural and functional disorders of genetic origin that will cause even more animal suffering.

Those animals that are subjected to genetic engineering to serve as "models" of various human disease conditions and as "tools" to test new diagnostic and treatment procedures will also suffer. And to what final end or purpose? Surely the betterment of humanity and social progress should not become even more dependent on the exploitation and suffering of other sentient, nonhuman beings.

Geneticist F. B. Hutt concludes that for *every disease* of farm animals that has been adequately investigated for evidence of a genetic basis, such evidence has been found.[6] And although selective breeding in poultry has been successful in reducing the incidence of some diseases, "breeding efforts ended when the problem could be resolved by vaccination or medication," according to poultry scientist W. Hartmann.[7] Hartmann emphasizes that the "need for maximum short-term improvement of economical efficiency has determined selection principles in commercial [farm animal breeding] programmes *which rarely left room for selection of viability*" (emphasis mine).

As a consequence, the genetic basis of disease problems in farm animals has long been neglected. According to the American Veterinary Medical Association:

> Genetic defects hit the livestock owner where it hurts—in the pocketbook. Kansas State University pathologist Horst Leipold is leading the research effort into the causes of diseases which are estimated to cost cattle breeders between $5–10 million annually. So far, scientists have identified 88 genetic defects in cattle. Leipold discovered 12 of those himself. He has received referrals (and defective animals) from most of the 50 states. ... The work is especially important because artificial insemination is so widely used in cattle. One bull with a genetic defect could, through artificial insemination, pass the defect on to as many as 70,000 calves.[8]

6　*Genetic Resistance to Disease in Domestic Animals* (London: Constable Comp. Ltd., 1978).

7　*World Poultry Science* 41(1985): 20–35.

8　American Veterinary Medical Association, *Animal Health News and Feature Tips* 1 (Summer 1985): 1.

According to geneticist Lawrence D. Young of the Roman L. Hruska U.S. Meat Animal Research Center in Clay Center, Nebraska, birth defects occur in at least 1 percent of all newborn pigs (*Acres USA*, May 1985). Although the causes for about 75 percent of these congenital abnormalities are unknown, a combination of environmental factors and genetic influences, especially related to inbreeding and selective breeding for high performance, is most likely involved.

As Lawrence Anderson has shown, many of the health problems that afflict livestock and poultry are related to selectively breeding for certain utility traits such as rapid rate of growth, which has been linked with calving difficulties and higher mortality rates.[9] Genetic engineering aimed at increasing the utility of farm animals is thus likely to intensify these already existing problems.

SUFFERING IN THE ABSENCE OF DISEASE. Biotechnologists' promise to use genetic engineering toward the improvement of health and disease resistance, as well as toward the utility of farm animals, thus actually reducing their overall suffering, is a falsehood of considerable magnitude. The new and profitable vaccines and other biologics that are being developed by the biotech industry will help reduce the incidence of certain diseases and associated *secondary suffering* in farm animals. The same can be said of genetically engineering the animals themselves to be resistant to specific diseases. But the stress and *primary suffering* that arise from the consequences of how they are selectively bred, raised, and handled and that have led to dependence on vaccines and other biologics to protect their weakened immune systems will not be eliminated.

---

[9]   *The Chance to Survive: Rare Breeds in a Changing World* (London: Cameron & Tayleur, 1978).

In fact, with such artificial supports (vaccines and drugs), *primary suffering will increase* as producers are able to adopt even more intensive methods of animal production. In other words, *the absence of actual disease does not mean an end to animal suffering* under current farm animal husbandry conditions. Using biotechnology to control infectious and contagious diseases will do little to improve the welfare of farm animals. In the final analysis, it may actually jeopardize their welfare even more.

Hormonal Manipulations

Injecting genetically engineered growth hormones, or splicing in growth hormone genes to make animals grow faster or produce more milk, essentially involves their transformation into "eating machines." A growth hormone gene transferred from rainbow trout into carp produced some fish that grew bigger and faster than normal. Researchers hope that such fish will keep eating and growing during winter months when most normal fish do little of either, according to *Science News* (133(1988): 374).

Considerable controversy has arisen over the use of injections of genetically engineered bovine growth hormone (or bovine somatotropin—BST) into dairy cows to make them more productive. Aside from consumer fears of hormone residues in milk and potentially disruptive effects on the dairy industry, particularly to smaller dairy operations, the additional stress to already high-yielding cows has caused much debate.

Only a few studies on a small number of dairy cows have been conducted, and not for any significant period. Veterinarian David S. Kronfeld notes: "Favorable responses to BST have been presented promptly, loudly and repeatedly. Unfavorable results have been delayed, subdued and obscured."[10] He has documented studies showing increased health problems, including mastitis (infected udders), reduced disease resistance, and reduced fertility, in BST-treated cows. Lameness is also more of a problem in BST-treated cows. Farmers (one-quarter of whom are predicted to be pushed out of business during the first three years of BST use) will require additional management skills to deal with these new problems and risks, but the one at greatest risk is the dairy cow herself.

Future Concerns

The genetic engineering of animals (along with other biotechnologies such as superovulation, in vitro fertilization, embryo transplantation, cloning, and the creation of chimeras—like the sheep with goats' heads) is a relatively recent development. This means that, at present, there is a total lack of evidence that the welfare of animals that are subjected to this technology can be guaranteed and that the coincidental and contingent suffering of these animals can be avoided. It is only presumed and promised by the biotech industry that animal welfare will not be placed in jeopardy and that "unnecessary" suffering will be avoided. But can these promises and presumptions be believed? There is already evidence of suffering in neo-genomic animals (animals with new genetic constructions).

It should be emphasized that although we may not like it as a culture, the suffering of billions of farm and laboratory animals is justified today on the utilitarian grounds of "unavoidable necessity," a justification that I

[10]   "Biologic and Economic Risks Associated with the Use of Bovine Somatotropins." *Journal of the American Veterinary Medical Association* 192(1988): 1693–1696.

reject and have recently challenged.[11] The suffering of millions more animals can be anticipated as new uses for animal life are discovered by the genetic engineers and marketed. And once society becomes economically, medically, and in other ways dependent on the creation and exploitation of these neogenomic animals, their suffering will also be justified on the grounds of "unavoidable necessity."

Alternatives

There are many nonanimal alternatives that are already available or could be developed so that animals need not be subjected to the risks of genetic engineering. For example, improved handling, transportation, and housing and husbandry practices are alternatives to developing genetically engineered disease-resistant animals. Bacteria and even plants can be engineered to produce insulin and other pharmaceuticals. The genetic engineering of farm animals to perform the same task should be questioned. Animals are sentient beings and can suffer; bacteria are not capable of suffering.

Transgenic mice have been bioengineered to secrete human tissue plasminogen activator in their milk, which helps remove blood clots from heart attack victims. Cows may be the next animals to produce useful pharmaceuticals for the nascent molecular farming—or "pharming"—industry. If there is no animal suffering following certain genetic changes, it is difficult to argue against the use of such animals as "protein factories." After all, they have long been exploited for products far less vital to human

---

[11] See M. W. Fox, *Inhumane Society: The American Way of Exploiting Animals* (New York: St. Martin's Press, 1990).

health—namely meat, hides, milk, and eggs. But does a history of exploitation establish an ethically valid precedent for continued and intensified exploitation? Such developments (which will never be stopped on ethical grounds) should be accepted only on the condition that the animals are kept under humane conditions that fully satisfy their behavioral and social needs. If they provide us with life-saving pharmaceuticals, we surely owe them no less.

However, this new form of animal exploitation can be seen as a form of "genetic parasitism," whereby animals have certain human genes spliced into their genetic matrix so that they produce biologics beneficial to humans with various genetically linked deficiency diseases and physiological anomalies such as hemophilia and Factor IX deficiency. It is ironic that animals should be so employed to help humankind compensate for its own genetic deterioration, which is in part attributable to the contamination of the environment and the food chain with industrial pollutants and agrichemical poisons, which can cause genetic damage.

Alternatives to genetically engineering animals as "models" for various human diseases should also be sought if the application of research findings to human patients is not primarily preventive in nature. Means of preventing genetic and developmental disorders in humans include such nonanimal alternatives as genetic screening and counseling and decontamination of environmental chemical pollutants that are teratogenic and mutagenic (i.e., cause developmental and genetic abnormalities), carcinogenic, or impair immunity.

Conclusion

Why do we need to genetically engineer farm animals to boost productivity, especially in these times of agricultural surpluses and chronic overproduction? Most of the health problems of farm animals are best addressed by making much-needed improvements in overall handling, transportation, housing, and husbandry. Are the risks and costs of potential and actual animal suffering worth the benefits, and who will be the primary beneficiaries? Certainly not the animals. The genetic engineering of laboratory animals is primarily the domain of profitable interventive human medicine. This is surely of less importance than public health and environmental and preventive medicine, which no amount of genetic engineering of animals can advance.

In the final analysis, are the public interest and the good of society really being served (biotech industry promises aside) by the genetic engineering of animals? Long-term social and environmental consequences need to be considered, as do the ethics of this new technology. Tighter and more appropriate animal-welfare regulations, and better guarantees of corporate responsibility for animals' well-being, are needed.

The demeaning perception of animals as "nothing but cells on the hoof" is one that many biotechnologists and others share. This is to be expected from those who see animals as essentially devoid of any intrinsic nature and inherent value independent of what they can do for us. A demeaning perception leads inevitably to a lack of respect and compassion.

This "mechanomorphic" perception of animals is a modern version of the philosophy of René Descartes who, in the seventeenth century, contended that animals are unfeeling machines. The fact that this perception of animals is still prevalent three centuries later points to a disturbing irony: In spite of our vastly increased knowledge of animal physiology, genetics, and power over life itself, there has been no concomitant increase in respect for life and for animals as sentient beings—at least among some

of those who now have power over the genes of life and in
whose hands the fate of creation now rests.

This observation should not be construed as an admoni-
tion or indictment of genetic engineering biotechnology,
but as a warning to all who have a deep respect for
animals and for the sanctity of beings and the future of
creation. This mechanomorphic perception of animals as
assemblies of cells and utilizable genetic resources is now
being reinforced and affirmed by genetic engineering
industry "biocrats."

An editorial in *The Christian Science Monitor* (3 June,
1988) entitled "Beware the 'New Creationism'" gave the
following insightful warning:

> As scientists, businessmen, and the public look at
> the promise of genetic engineering, they need to
> constantly guard against a subtle and dangerous
> attitude. It could be called the new creationism.
>
> Its impact on human advancement can be just
> as detrimental as the "old creationism." This holds
> that a Supreme Being was responsible for creating
> the material universe and a hierarchy of distinct
> forms of life in a very brief span of time several
> thousand years ago.
>
> The new creationism replaces the Deity with
> human beings wielding the tools of molecular
> biology. Though focusing solely on organic life,
> the new creationism assumes responsibility for
> "creating" improved plants and animals strictly
> for human benefit in a fairly short time.
>
> In essence, new creationism threatens to erode
> humanity's respect for the intrinsic individuality
> and value of all forms of life. . . .
>
> As subtle forms of new creationism take hold
> on thought, they can lead to the kind of hubris
> reflected in the summary of a recent genetic-
> engineering article: "Regulators and the public are
> inhibiting a major new industry. They must better

understand the potentials and limits of biotechnology, and then get out of the way."

Get out of the way, indeed—as if those outside a
tight circle of experts have nothing of value to say
or no concerns to raise.

There have been several new developments in the
genetic engineering of animals that show how this new
industry is applying biotechnology in agriculture and medicine. How appropriate these new developments are—in
terms of real progress in improving agricultural practices
and human health—remains to be seen. The preceding
examples clearly reveal that the new creation and new
world order of the biotechnology industry are far from the
utopian dream of a world made perfect for humankind.
One can read between the lines of the new patent applications, news releases, and scientific reports concerning the
latest feats of genetic engineering and glimpse into
the future. The wonderworld of the new creationism is not
quite here today, but it may be upon us sooner than we
expect. A whole new generation of genetically engineered
animals is on the horizon. In the world of trade and
commerce, they will be regarded as "new" species—unique,
patentable commodities of the new world order—a subject
that is discussed in the next chapter.

THE
PATENTING
OF ANIMALS

I am not willing to let the marketplace
determine the future of the animal kingdom.
—REP. CHARLIE ROSE (D-N.C.), statement
before the Committee on the Judiciary,
U.S. House of Representatives, 22 July, 1987

According to *Science* magazine:

> A ruling by Board of Patent Appeals and
> Interferences of the United States Patent
> and Trademark Office [on 3 April, 1987] appears to
> have cleared the way for the patenting of animals
> with unique, man-made characteristics that do not
> occur in nature. . . . Until now, the patents have
> been granted to plants and microorganisms, but
> not to higher life forms. (10 April 1987, p. 144)

Thus the board broadened the interpretation of the 1980 Supreme Court decision in *Diamond v. Chakrabarty* (which was that genetically modified microorganisms can be patented) to include all life forms. The decision to permit the patenting of genetically altered animals (without congressional approval) reflects the arrogant presumption of authority by the U.S. Patent and Trademark Office to actually grant and protect the exclusive right to produce and/or sell a particular kind of animal—one that has been genetically altered.

It is the height of human arrogance to regard genetically modified animals as patentable by their creator-owners. They are not, as the Patent Office contends, entirely unique, "new" life forms, because no scientist has yet been able to create life in the laboratory. They are simply modifications of existing life forms, whose future is now imperiled on many fronts. The animal kingdom is "ours" insofar as it is in our common trust. According to Patent Office official Charles E. Van Horn, the Supreme Court ruled in the *Chakrabarty* case that Congress intended that "anything under the sun that is made by man" could be patented (*Science*, 10 April, 1987, p. 144). According to attorney Andrew Kimbrell, this legal interpretation is highly questionable, since he original statement from the Court was that "anything under the sun that is made by man *that is a new manufacture*" could be patented. A genetically altered animal is not a new manufacture.

What were the legal precedents, if any, and preexisting perceptions at the time of the board's ruling on 3 April 1987 that animals can be patented? Apparently, the legal precedents were tenuous at best. It was only by a margin of one vote that the U.S. Supreme Court (in the 1980 *Chakrabarty* decision) ruled that microorganisms could be patented. Bacteria and viruses are not as sentient as animals—being closer to chemical processors and plants—and it is likely that if a transgenic animal had been involved, such as a rhesus monkey with human genes, the Court would not have given the okay and opened the door to the patenting of all life. The other precedent for the patenting of animals, which is not a precedent at all, is the Plant Variety Protection Act. Animals are not plants, and it is both legal fiction and biological nonsense to put animals like dogs, chimpanzees, and horses in the same category of patentable inventions as genetically engineered microbes and plants or microwave ovens and Velcro fasteners. Patent law was originally intended to cover only nonnaturally occurring manufactures and compositions of matter: It was for inanimate, nonreproducing, nonnatural creations.

What of selectively bred or purebred varieties of animals? These cannot be patented. The European Patent Convention clearly prevents organizations from claiming sole rights over a new breed or variety of animal (although steps are being taken by the European Economic Community [EEC] to change this convention to allow for the patenting of all genetically engineered life forms, which are referred to as "self-replicable matter").[1] Since such animals do not exist under natural conditions, and since transgenic animals are not entirely new creations but are varieties of already existing breeds and species, it is illogical that genetically engineered animals should be patentable.

The law is surely based on objective reason and those valued human sensibilities of compassion, respect, and justice. The only reason to endorse the patenting of animals is financial. Subjective economic concerns—which some call greed—should not take precedence over reason,

---

[1]   The European Patent Office recently rejected its policy that genetically engineered animals cannot be patented, ruling that the E. I. du Pont de Nemours & Co. "oncomouse" could be patented. However, the European Community has not yet taken a definitive stand on this issue.

compassion, respect, and justice, which the law is intended to serve for the good of society. When the law instead serves the financial interests and economic imperatives of the technocracy and is entranced by the promises of science, we have an institution that can no longer be trusted to serve the public interest and the common good. Technology should not determine and control human choice. Rather, human choice should control technology.

The patenting of animals will have profound worldwide socioeconomic consequences, forecasts of which do not bode well for farmers either here or abroad. A ban on animal patenting would not make U.S. agriculture less productive, as some contend. It is already blighted by chronic overproduction. And how would farmers be monitored to pay user fees to holders of animal patents?

Some contend that a ban on animal patenting would arrest medical progress. This is not so. Medical progress has been significant without animal patenting for many decades. Furthermore, the collective knowledge of society that has led to the creation of transgenic animals has been supported by the culture as a whole since the beginning of civilization. It is surely a violation of the common trust to privatize the fruits of such knowledge by now extending patent protection to cover the whole animal kingdom and the creative process itself.

The proponents of genetic engineering and the patenting of animals cite the use of selective breeding and crossbreeding in the domestication of animals to support their case. The term *domestication* means to tame, "to accustom to home life." The domestication of animals by means of selective breeding and socialization with humans in early life is very different from their wholesale industrial exploitation through genetic engineering and patenting.

Domesticated animals have been subjected to genetic alteration for thousands of years. Contrast the structure and physiology of a Holstein cow and a Hereford bull or a Saint Bernard dog and a Mexican Chihuahua. To now permit the patenting of animals subjected to genetic alteration, principally by means of genetic engineering, could have several adverse consequences. From a scientific perspective, these include the following concerns.

1. *No regulation.* The floodgates will be opened wide once genetic engineering research on animals is patent protected, because biotech companies will have the protection they need to secure a monopoly over new "intellectual property" (i.e., genetically engineered animals). This will mean a dramatic increase in animal experimentation for agricultural, biomedical, and other industrial purposes, which cannot be effectively regulated. The outcome of many genetic experiments cannot be predicted in relation to the animals' health and welfare or in relation to the long-term social, economic, and environmental impact. In many instances animals will be abnormal at birth, and generations will suffer until techniques are perfected and accidents prevented.

2. *Monopoly.* Patenting could result in monopoly of genetic stock and predominance of certain genetic lines of animals over others, with an ultimate loss of genetic diversity within species. This could have a significant impact on agriculture as well as adverse social, economic, and ecological consequences. And if farm animals are patented, will farmers have to pay a user fee for offspring and crossbreeds, and how would this be enforced?

Lynn McAnelly, a technology analyst with the Texas Department of Agriculture, Austin, sent out over 1,700 letters to livestock producers in Texas, asking them whether patenting animals would increase or decrease their costs. Of about 500 responses, 96 percent predicted that costs would go up. The consensus was that only patent holders and large agribusinesses would benefit. It was also felt that animal patenting would provide increased opportunity for large corporations and syndicates to gain control of the industry and that any cost advantages of patented animals would wind up in the pockets of big agribusiness. As a result of the survey, Texas Agriculture Commissioner Jim Hightower asked the Texas congressional delegation in Washington to support a patent-moratorium bill.

3. *Effect on wildlife.* Patenting would also cloud the ownership of wild animals. In the United States, wildlife is held by governments, both state and federal, as a common public trust. If a deer is altered genetically, can it be patented? Will people be disenfranchised of ownership of their wild animals? America's wildlife is far too precious to get caught or lost in a discussion about patenting and ownership. The American people own wildlife, to the extent that anyone does, and patenting would mean a very real threat to such ownership.

The biomedical industry will play upon public fear to block all attempts to prohibit the patenting of animals. It will tell us that the march of modern medicine will stop dead in its tracks without patent protection. The fact remains that medical advances have been made in the past without the patenting of genetic engineering techniques and of animal models. And we should recognize that patenting in this area could actually inhibit medical progress since, for proprietary reasons, research findings of privately funded laboratories and university research institutions would not be shared. There would also be considerable unnecessary and costly duplication of research, because the patenting of animal models would encourage a competitive, rather than a collaborative, research atmosphere, to the ultimate detriment of the public's best interests.[2]

2   For details see M. Kenney, *Biotechnology: The University-Industrial Complex* (New Haven, CT: Yale University Press, 1986).

Patent Ethics

From an ethical perspective, the patenting of animals reflects a cultural attitude toward other living creatures that is contrary to the concept of the sanctity of being and the recognition of the interconnectedness of all life. The patenting of life reveals a dominionistic and materialistic attitude toward living beings that denies any recognition of their inherent nature.

Left unopposed, the patenting of animals will mean the public endorsement of the wholesale exploitation of the animal kingdom for purely human ends. Since humans are also animals, then logically there should be no legal constraints on the patenting of techniques to genetically alter human beings for the benefit of society. But there are ethical constraints (as well as the Thirteenth Amendment to the U.S. Constitution, which prohibits the ownership of one person by another) that protect the sanctity and dignity of human life. To permit the patenting of animals will effectively eliminate ethical constraints on genetically altering other animals, and eventually humans, for the purported benefit of society. Such a utilitarian attitude toward life is a reflection of the ethical blindness of the times.

Supporters of animal patenting have argued that if the patenting of animals is prohibited, companies engaged in the genetic engineering of animals will fall back on trade secrecy to protect their investments in research and development. The Trade Secrecy Act, they reason, would make it difficult for those concerned about animal welfare to know what had been done to genetically engineered animals. If patenting were approved, all details would be available to the public in the patent application. In reality, however, public access to such information would be of little help in protecting animals' rights and welfare. With or without the patenting of animals, the genetic engineering of animals is being done. And by the time a patent application is filed, all the research on the animals has already been completed. Thus rigorous ethical guidelines concerning the welfare of animals subjected to genetic engineering are needed before the onset of new research projects. Knowing what has happened to them after a patent has been granted is of little avail.

It should also be remembered that the U.S. Patent and Trademark Office does not, as a rule, consider the ethical, moral, and social consequences of patent applications. The essentially amoral and objective role of this governmental agency is dramatically illustrated by the granting of patent number 4,666,425 to attorney and engineer Chet Fleming for his "discorporation" life-support system. This system, which Mr. Fleming has never actually used, is designed to keep the isolated head of an animal alive. This patent application was apparently filed to provoke greater concern for the future of new technologies and for their moral, ethical, and social consequences, which Mr. Fleming requested the Patent Office to consider in his application. But apparently, it did not. The office granted him a patent without further question.

The primary reason for the patenting of genetically engineered animals is to protect private interests, and opposition to animal patenting is clearly a threat to the biotechnology industry. Animal patenting is an issue quite distinct from genetic engineering per se. It is an issue that is linked with private interests and monopoly on the one hand, and the public endorsement of animals as patentable commodities and inventions on the other. As such, the patenting of animals is an ethical issue, supported primarily by economic concerns and an attitude toward nonhuman creatures that is contrary to the mainstream cultural traditions of reverence for life and respect for animals and the natural world. Patent protection will do nothing to protect the rights and welfare of animals and will serve to further undermine those cultural traditions that opponents of animal patenting value so highly.

Chronology of Animal Patenting

- On 7 April, 1987, the U.S. Patent Office interpreted patent law to allow for future patents on animals changed or altered through genetic engineering or similar techniques. Relying on the Supreme Court decision in *Diamond v. Chakrabarty*, 447 U.S. 303 (1980), which held that microorganisms could be patented, the Patent Office determined that such genetically altered animals were nonnaturally occurring "manufactures" and "compositions of matter" and thus could be included under section 101 of the Patent Act as patentable subject matter.
- The Supreme Court decision made no mention of animals, and Congress has never approved the patenting of living things except for certain specified plants in legislation passed in 1930 and in 1970.
- On 17 April, the Humane Society of the United States (HSUS), the Foundation on Economic Trends, and a coalition of animal-welfare organizations representing five million people petitioned the Patent Office to rescind its controversial decision. The coalition included 11 national farm groups, 24 religious leaders, 21 animal-welfare organizations, and 8 environmental and public interest groups. It was concerned about long-term

ethical, animal-suffering, environmental, economic, and governmental consequences of the patenting of animals.

- In 1987, the Senate passed a Hatfield (R-Ore.) amendment to the continuing resolution, which would have temporarily blocked patenting. But the amendment was dropped in conference when Commissioner Donald Quigg stated that the Patent Office would not be able to act on animal patent applications before the end of the fiscal year on 30 September 1987. Patent Office officials later stated that they might be able to issue patents as soon as 1 April 1988.

- Chairman Robert Kastenmeier (D-Wis.) of the House Judiciary Subcommittee on Courts, Civil Liberties, and the Administration of Justice held a series of four hearings on the issue. No further action was taken. John Hoyt, president of the HSUS, testified on 11 June, 1987, stating that patenting is "inappropriate, violates the basic ethical precepts of civilized society and unleashes the potential for uncontrollable and unjustified animal suffering."

- On 5 August 1987, Rep. Charlie Rose (D-N.C.) introduced H.R. 3119 to impose a moratorium on the patenting of animals so that the potential adverse implications of such patenting could be carefully studied. The HSUS and others wrote to Rose pledging to conduct studies during the moratorium period. On 29 February, 1988, Sen. Mark Hatfield introduced a moratorium bill, S. 2111, in the Senate.

- On 13 April 1988, the U.S. Patent Office issued the first patent on a genetically engineered animal. Harvard University researchers had developed the "oncomouse," a genetically engineered, cancer-prone mouse. Funding for this research came from DuPont Chemical Co., one of the world's major producers of carcinogenic chemicals.

- Rep. Kastenmeier drafted legislation to make patent-user exemptions for family farmers and scientists in order to quell some of the increasing public opposition. Meanwhile, steps were taken in Europe by the European Economic Community to change existing laws that prohibited the patenting of selectively bred plant and animal varieties so that all genetically engineered life forms might be patented.

- On 13 July 1988, under pressure from the State Department of Commerce, which insisted that a moratorium on animal patenting would harm U.S. industrial competitiveness, Rep. Kastenmeier's subcommittee voted 8 to 6 against Rep. Rose's moratorium. This established the United States as the first nation to officially endorse the patenting of all life forms subjected to genetic engineering.

- On 2 August 1988, the House Judiciary Committee approved Rep. Kastenmeier's legislation on animal patenting, which would have exempted farmers from paying royalties on the offspring of patented transgenic animals. According to opponents in the biotech industry this would have removed many economic incentives for developing genetically engineered animals. Supporters of the exemption saw it as vital to the survival of family farms. (It never became law, most likely as a result of pressure from the inner circle that is now the Council on Competitiveness.)

- Since the patenting of the oncomouse in the United States there have been no further animal patents awarded.

- Some 145 animal patent applications are now awaiting approval at the U.S. Patent and Trademark Office. Approximately 80 percent of these have medical utility, and the remainder involve agricultural animals.
- A new bill was introduced in the Senate (S. 1291) by Sen. Hatfield on 13 June, 1991 to impose a five-year moratorium on the granting of patents on invertebrate and vertebrate animals, including those that have been genetically engineered. I supported this bill with the following statement published in the *Congressional Record* on that day (pp. 7818–7819).

In order for society to reap the full benefits of advances in genetic engineering biotechnology, the social, economic, environmental, and ethical ramifications and consequences of such advances need to be fully assessed. Considering the rapid pace of developments in this field, which will be spurred on by the granting of patents on genetically altered animals, a 5-year moratorium on the granting of such patents is a wise and necessary decision. A moratorium will enable Congress to fully assess, consider, and respond to the economic, environmental, and ethical issues raised by the patenting of such animals and in the process, establish the United States as the world leader in the safe, appropriate, and ethical applications of genetic engineering biotechnology for the benefit of society and for generations to come.

- On 28, April 1992, Benjamin Cardin (D-Md.) introduced a bill (H.R. 4989) in the House of Representatives also calling for a 5-year moratorium on granting patents on invertebrate and vertebrate animals, inlcuding those that have been genetically engineered.

It is very likely that the White House Council on Competitiveness, chaired by Vice President Dan Quayle, will attempt to block these bills. This same Council has been actively working to deregulate the entire biotechnology industry. Its proposed administrative and regulatory guidelines for the Environmental Protection Agency and U.S. Department of Agriculture are such that the social, economic, ecological, environmental, and animal-welfare risks, costs, and consequences of new developments in biotechnology will be virtually ignored.

Clearly, although the genetic engineering of animals is not likely to be stopped, increased public awareness and censure of the biotechnology industry and its political allies are essential. A five-year moratorium on the patenting of "new" animal creations would be prudent and timely. We are moving into a new world order of free trade, which should be conditional upon effective international regulations and the adoption of the most stringent controls and regulations over biotechnology by all nations. Otherwise, the privatization of the world's resources, and of the genetic material of life itself, coupled with the misapplication of genetic engineering biotechnology in agriculture and medicine, will be against the public interest and the good of generations to come.

The End of Nature?

The industrialized commercial exploitation of life must be examined in the much broader context of what we are doing to the natural world and to the created order. Some see the Earth being turned into a desecrated and polluted wasteland through the synergism of the desperate poverty of the many and the insatiable greed of the few. The application of genetic engineering biotechnology under man's self-centered mind-set will accelerate the now unnatural pace and direction of evolution on this planet and, thus, guarantee the end of the natural world if this trend is not confronted and changed.

Genetic engineering and the patenting of animals herald a very different attitude about and relationship with animals than ever before. Biotechnology has placed us at the threshold of gaining absolute control over the life process itself. It is as imprudent to assume a neo-Luddite (all technology is bad) attitude, and vainly seek the total prohibition of the inevitable, as it is to become entranced by the utopian promises of biotechnology and oppose any questioning or public debate on the issue. The scientific "priesthood" of the technocracy is neither evil nor infallible. What is urgently needed is a more open public forum on this issue and legislation to prohibit the patenting of all life until scientific serendipity, the profit motive, and our ethical sensibilities are given equal and fair consideration.

It has been said that where there is no vision, the people shall perish. Those who have the vision of a future utopia through genetic engineering—and there are many, considering that this was the biggest growth and investment industry of the 1980s—may be suffering from what theologian Thomas Berry calls "technological entrancement." As he sees it, this entrancement leads us to recreate the world in our own image to serve all our needs, no matter how spurious, rather than "reinventing" ourselves to assume a more creative planetary role. As history teaches us, the consequences can be highly destructive and even injurious.

Although some may not see the advent of genetic engineering and the patenting of life as apocalyptic, or even as heralding a future utopia, it would be wise for us all to find the middle ground between the extremes of probability and delusion. That middle ground is not simply to condone and regulate all genetic engineering research on living beings, but to consider each project on a case-by-case basis. (The importance of corporate responsibility in this regard is considered in Chapter 8.)

The privatization of genetic engineering of animals through the patent process should be prohibited by Congress. The greater good can be assured only if we, as a society, continue to hold to the principles of democracy. In the context of genetic engineering, this amounts to a trans-species democracy: respect for the rights and sanctity of all sentient beings.

Postscript: Patenting Human Genes

If the rush to patent new genetically engineered life forms is not evidence enough of monopoly and greed, then the recent disclosure that the National Institutes of Health (NIH) are trying to corner the market on *human* genes should serve as an alarm to us all. According to reporter Larry Thompson, NIH has applied for 347 patents on human genes and may soon file applications on 2,000 more.[3] This could lead to a highly competitive international "patent gold rush" to gain exclusive rights to use certain human genes to manufacture various products, primarily pharmaceuticals. The precedent that this sets has enormous ethical and economic ramifications, yet NIH Director Bernadine Healy is quoted by Thompson as saying, "I think the discussion has become a little silly because it has been elevated to the level of some moral, science policy, [an] ethical issue." But is it not just that? A few human genes have already been patented, but only in a specially prepared form used to manufacture gene-based

[3] "NIH Rush to Patent Human Genes," *Washington Post*, 28 October 1991, p. A3.

drugs. No patents have ever been submitted or issued on the scale proposed by NIH. Such large-scale patenting would be a disincentive for small U.S. biotech companies that could not afford to risk venture capital developing a new product based on a gene only to discover that it has already been patented by NIH.

Addendum

According to reporter Alex Barnum (*San Francisco Chronicle*, May 7th 1992), the U.S. Patent and Trademark Office has informed GenPharm International of Mountain View, California, that it will soon receive a patent on a new mouse they have developed called the TIM mouse for *transgenic immunodeficient mouse*. The genes that code key parts of the animal's immune system are destroyed so that the mouse develops without an immune system and produces offspring with similar defects. The TIM mouse will be given human immune cells and serve as a living laboratory to study AIDS and other immune-system diseases.

CHAPTER 7

NATURE-
CONSERVATION,
BIODIVERSITY
AND BIOTECHNOLOGY

*The glory of the human has become the desolation of
the earth. This I would consider an appropriate way
to summarize the twentieth century.*
—THOMAS BERRY, Riverdale Center for
Religious Research, New York

*T*he worldwide impact on wildlife and wildlands will
be devastating if innovative applications of biotechnology are not initiated with an environmentally sound
philosophy. For example, farm animals may be modified
genetically to enable them to adapt to previously un-
suitable tropical, arid, wetland, and other habitats. They
will be made resistant to various diseases (such as sleeping
sickness) to which they were previously susceptible; diseases to which local wild life have often developed a *natural*
resistance. The introduction of these genetically modified

livestock will escalate the rate of destruction of wildlife habitat, accelerate the decline in biodiversity, and lead to the displacement and ultimate extinction of indigenous wildlife species.

As ecologist Daniel H. Janzen observes:

> Tropical wildlands and most of the earth's contemporary species still exist because humanity has not had organisms capable of converting all tropical land surfaces to profitable agriculture and animal husbandry. Within one to three decades, organisms modified through genetic engineering will be capable of making agriculture or animal husbandry, or both, profitable on virtually any tropical land surface. Agricultural inviability, the single greatest tropical conservation force, will be gone. . . .
>
> An enormous amount of wildland genetic information will be obliterated overnight. And it is precisely this diverse and exotic genetic information that will be most eagerly sought by the genetic engineering industry once we are past the stage of simply making better beef, beans, and corn. It is very much in the selfish interests of this growing industry to join forces with the conservation community. (*Science* 236[1987]: 1159)

A Transformed Biosystem

Patented living bacteria and viruses have already been released into the environment. They will soon be followed by genetically engineered fungi, nematodes, insects, and other plant parasites developed to protect crops from disease. The Environmental Protection Agency (EPA) and the U.S. Department of Agriculture's (USDA) Animal and Plant Health Inspection Service (APHIS) are in charge of regulating their release under regulations that are still in the process of being formulated. The entire complex ecology of Earth may rapidly be transformed into an industrial biosystem. Already several hundred field-test releases of genetically engineered organisms have been approved by U.S. authorities and by foreign governments (see Table 7.1 and Figure 7.1).

This next round of pesticides and fertilizers from genetically engineered pest-killing and nitrogen-fixing bacteria, along with crops (including trees) that produce their own pesticides, herbicide- and disease-resistant seeds and feed-efficient, disease-resistant livestock, will cause the displacement and extinction of wild plant and animal species. Ecological shifts will escalate the trend toward global bioindustrialization as various recombinant varieties of organisms, from microorganisms to plants and livestock, are introduced. The few tropical forests and other wildlands that remain in the biotech world of the future will have to be protected as valuable genetic "banks" for the pharmaceutical and chemical industries. It is vital that these areas be saved from loggers, dam builders, monoculture foresters, and cattle and sheep ranchers, since recent studies have shown that large areas of natural habitat must be preserved in order to prevent a decline in species diversity.

In addition to the displacement of wildlife and destruction of natural habitats by the expansion of agribiotechnologies, the long-term rebound effects on remaining wildlife populations and habitats need to be considered. For example, given a large acreage of herbicide- and disease-resistant trees or soybeans, what happens to the insects and their predators (other insects and birds) that are displaced? They may be forced to seek new niches and compete in other ways with indigenous creatures in adjacent nonagricultural, recreational, wildlife-park, or wilderness land. This rebound effect could be further aggravated by the aerial dissemination of sprayed bacterial

FIGURE 7.1.

STATES WHERE RELEASES OF
GENETICALLY ENGINEERED ORGANISMS HAVE BEEN APPROVED

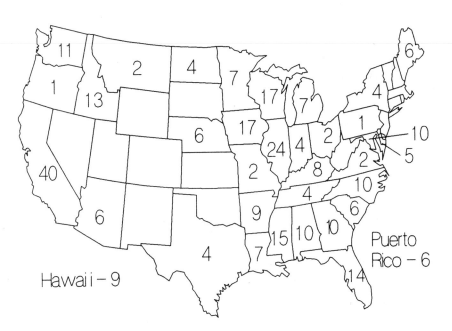

Total Tests - 303

THE NUMBER IN EACH STATE EQUALS THE NUMBER OF
TESTS APPROVED BY USDA & EPA IN THAT STATE AS OF
11/21/91. WE ASSUME THAT APPROVED TESTS HAVE BEEN
CONDUCTED.

Source: National Wildlife Federation, 1991, adapted from data provided by U.S. Department of
Agriculture and U.S. Environmental Protection Agency.

TABLE   7.1.
___

FIELD TESTS, BY COUNTRY
(Summer 1990)

United States	93
European Community	62*
Canada	18
Australia	4
Argentina	1
Japan	1
Other	8

Source: U.S. Department of Agriculture, 1991.
Note: Because of differences in definitions, some of these statistics for countries
   outside the United States may include tests of modified microorganisms as
   well as transgenic plants, but these tests are relatively few.
*28 in France; 12 in Belgium.

pesticides onto nonagricultural land. And it would be imprudent to engineer agricultural crops and forest-industry
trees to be resistant to air pollutants and ultraviolet radiation but ignore the health and environmental consequences
that could further devastate wildlife and their habitats.

Consider the recent award of almost $2 million in
public funds by the U.S. Army Medical Research Institute
of Infectious Diseases at Fort Detrick, Maryland, to
Molecular Genetics Inc., a genetic engineering company in
Minnetonka, Minnesota. Molecular Genetics is to develop
a recombinant DNA vaccine for Rift Valley fever virus, a
disease of cattle and humans prevalent in the Middle East
and Africa. Although it would perhaps be inhumane to
oppose research to prevent this disease, the impact on the
environment and wildlife is a legitimate concern. This is a
Catch-22 situation: The increase in human population
following the development of a successful Rift Valley fever
vaccine will create the need for an expanded livestock
population to sustain those people. Without rigorous

birth-control programs and the adoption of alternative agricultural and food habits, a vicious circle will develop.

In most livestock disease-control and eradication programs in the third world, the goal of increased livestock production is not to provide food for local consumption but to export it to more affluent countries, especially the European fast-food hamburger market. Profits generally benefit the powerful few in these third-world countries, where malnutrition and environmental degradation (especially desertification) are increasing and will not be rectified by raising more cattle as a cash crop for export.

Agribiotechnology and Sustainability

One of the big selling points of the new biotechnologies in animal agriculture and veterinary medicine that cannot go unchallenged is that by making farm animals more productive, fewer farm animals will be needed to satisfy market needs. This is supposedly consonant with the adoption of sustainable agricultural practices, since it will help reduce the negative environmental impacts and economic inefficiencies of raising less healthy and thus less productive farm animals for human consumption.

This rationalization for the adoption of new biotech products is reinforced by the assertion that these products are safer for consumers and the environment than many of the pesticides, antibiotics, and other products of the petrochemical-pharmaceutical industrial complex (PPIC). It is ironic that a decade ago the PPIC insisted that its pesticides and antibiotics were safe; now it is saying that there are *safer* alternatives.

What of the adverse environmental impact of animal agriculture in the developed world? The PPIC has profited from and contributed to the adverse environmental impact of a nonsustainable animal agriculture for decades. The high livestock populations in the United States and in many European countries have created serious environmental problems, especially in waste management. These problems will worsen if the demand for animal protein by an ever-increasing populace is not drastically reduced and if livestock production practices are not integrated with more sustainable agricultural systems.

According to Dr. Lester Brown of Worldwatch Institute, 600 million tons of grain—more than one-third of world production—are fed to livestock. This wasteful misuse of grains must be curtailed, since Worldwatch Institute, among others, has clearly documented that despite increased food production as a result of new agritechnologies over the past three or four decades, the unsustainable overuse of land and water worldwide will mean serious food shortages in the near future. Other than a radical change in farming practices toward sustainability, there are no foreseeable or cost-effective technological solutions to prevent this pending global catastrophe. The application of new biotechnologies in the development of sustainable agricultural practices is being delayed by the PPIC because of its financial and ideological commitment to nonsustainable, overcapitalized livestock and crop production systems.

It should be recognized that most agribiotechnology innovations to date have been aimed at lowering the costs of farming rather than increasing overall productivity. The food surpluses and low prices of the 1980s made anything that promised to cut costs and increase profit margins very appealing to farmers. Hence the primary focus of agribiotechnology has been to improve the efficiency and profit margins of current farm animal and crop production practices and thus maintain the status quo of a fundamentally nonsustainable and capital-intensive agribusiness industry.

Investors in animal agriculture biotechnology should think twice. Today's human population of 5.3 billion is expected to double within the next generation. It will become increasingly unethical for richer countries to squander their resources and exploit poorer nations to raise feed for farm animals, since meat and milk production is a highly inefficient way to make food and no way to feed a hungry world.

It is a matter of public record that the PPIC is heavily invested in the research and development of new biotechnologies. Because of its global monopoly, the PPIC has set great marketing hopes on these new products, beginning with bovine growth hormone. Genetically engineered vaccines and disease-diagnostic kits are already being widely marketed to prop up intensive animal agriculture in the United States and other industrialized nations. But it is in the third world where such applications are most urgently needed, where millions upon millions of livestock are riddled with disease, making them suffer and greatly limiting their productivity and utility.

Biodiversity

Another ploy of the biotech industry is to promote the idea that genetic engineering is good for all life because it can help increase the Earth's biodiversity. This notion is patently absurd. A biotechnology-supported livestock industry that expands to meet the public demand (and expectation) for meat and dairy products as dietary staples, coupled with more people on the planet, will lead to a loss of natural biodiversity as more and more wildlands and wildlife are obliterated. More trees will be felled, more swamps drained and more dams built to convert more land into pasture and cropland, primarily to feed livestock. Thirty-seven percent of the land in the United States is used for meat and milk production. Conservationists, environmentalists, agri-economists, and climatologists are now joining to question the environmental and socioeconomic impact of the livestock industry, which is now almost wholly a subsidiary of the PPIC.

The same petrochemical-pharmaceutical industrial complex (with its mining, energy, forestry, livestock, and agribusiness subsidiaries) that is responsible for much of the recent loss of the Earth's biodiversity now hopes to profit by making its next venture—the genetically engineered, industrialized exploitation of the biosphere— politically and publicly acceptable. Natural biodiversity will be drastically reduced if there is international political and public acceptance of genetic engineering and of the patenting of "new" corporate-owned and -created animals, plants, microorganisms, and products.

There is also an accelerating loss of biodiversity caused by agribusiness's overreliance on a few utility strains and varieties of seed stock and livestock. It is these commercial strains that are now being genetically engineered and patented, while rare breeds of farm animals and plant seed stocks are not being adequately conserved. Yet it is the latter that are a vital genetic resource for the future. Their loss will greatly limit the flexibility of agriculture to adapt to climatic change, and it will severely curtail the adoption of bioregionally sustainable and ecologically sound agricultural practices.

The history of the biotech industry clearly reveals an intensifying race for corporate control over the world's potentially useful germ plasm.[1] This will inevitably lead to

---

[1]   For documentation, see Jack Doyle, *The Altered Harvest* (New York: Viking Press, 1987), and Pat Roy Mooney, *The Law of the Seed. Another Development and Plant Genetic Resources Development Dialogue* (Uppsala: Dag Hammerskjold Foundation, 1983), pp. 1–2.

a loss of rare animal and plant breeds, strains, and species. It could also mean a loss of cultural diversity as more and more bioregional communities and nations, especially in the third world, are co-opted and coerced into the second industrial revolution of genetic engineering biotechnology. The result will be a new wave of colonialism and a new world order whose power base is the corporate boardrooms of the industries that have a monopoly over the genes of life. Decisions made in these boardrooms will determine how people live, what varieties of crops and animals they raise, what they eat, and even how they value life itself. Concern for this potential threat to cultural diversity has been voiced by highly credible agencies and analysts.[2] In sum, an informed public voice is demanding corporate reponsibility and accountability in its applications and claimed benefits of this new genetic technology. It should benefit the environment and the populace first and equally; corporate interests and investments should be consonant with this altruistic ideal and attainable possibility.

Advances in animal agriculture biotechnology that are not integrated with a humane and sustainable agricultural paradigm that is socially just and "environmentally friendly" will be opposed by an increasingly vociferous, influential, and informed public. Opposition is also growing out of new global alliances that the often shortsighted, if not unethical, activities of the PPIC have indirectly helped create—as between the disenfranchised indigenous peoples of the third world and concerned consumers of the industrial world.

[2] See Cary Fowler et al., *The Laws of Life. Another Development and New Biotechnologies Development Dialogue* (Uppsala: Dag Hammerskjold Foundation, 1988), pp. 1–2.

The Potential Threat

From the above analysis it is clear that the biotechnology industry is potentially one of the most serious threats to the biodiversity and ecological integrity of planet Earth. The threat will become a reality if this technology is applied with the same values and attitude toward life and the biosphere that sanctioned and promoted the wholesale application of pesticides and the development of capital-intensive monoculture farming and forestry. *But this is not to say that this new technology could not be used appropriately.* For example, it could be used to engineer plants to help halt the spread of deserts; to develop micro-organisms and plants to synthesize essential biologics, such as insulin and antibodies; to help in water treatment (so-called bioremediation) to remove pesticides, heavy metals, and other industrial and agrichemical poisons.

So far, the risks to wildlife populations and global biodiversity far outweigh any short-term benefits that recombinant DNA biotechnology can promise. But this need not be the case. The biotechnology industry has the power and responsibility to initiate ecologically appropriate programs and policies and to act creatively for the sake of all life on Earth. This entails not simply respecting the natural world or the sanctity of being, but realizing that it is economically prudent in the long run, as history has taught us, to preserve the integrity and diversity of biotic communities.

Botanist Peter Raven underscored the importance of conservation for the biotechnology industry in an interview in *Genetic Engineering News* (July/August 1988, p. 29):

> Biotechnology and genetic engineering companies must realize that both progress and profitability

depend on their ability to understand and manipulate biological diversity. The process of extinction, however, is eliminating an enormous array of possibilities, and biological diversity is literally slipping through our hands. . . .

Human activities, especially in tropical and subtropical forests, are responsible for the destruction; rapidly growing populations, extensive poverty and a lack of knowledge about sustainable agriculture and forestry are the main causes. Vegetation throughout the world is being destroyed to grow crops, without any firm agricultural policy in mind. Whole forests are cut down to supply firewood for daily activities. Large numbers of species of plants, animals and microorganisms are lost in this wholesale deforestation.

Dr. Raven concluded that he does not look for the human race to go on mindlessly increasing to the point where everything is destroyed:

I look for it to become stabilized in such a rational and sentient way that there can be room for conservation of certain pieces of real estate in between that. That's why I rejoice in the application of biotechnology and hybridization and improvement of all kinds of domesticated crops and animals and everything else—because we have a very serious problem. We are managing, wasting, using, consuming the entire productive capability of the planet that we live on, and unless we get very serious about that and accept it as an immediate task that we've got to address, we simply don't have any hope for survival.

We must be extremely cautious about releasing genetically engineered organisms into the environment. The deliberate and accidental release of exotic, nonindigenous plant and animal species has caused considerable harm already, which should not be ignored by the biotechnology industry. Those who forget history are condemned to repeat it.

It is a matter of public record that the biotechnology industry has held firm to the belief that existing government regulations circa 1980, under the aegis of the FDA, EPA, and USDA, were fully adequate and appropriate in assuring the safety and efficacy of their products; and that animal-welfare, environmental, and socioeconomic concerns had no foundation and were voiced by a minority faction of antiestablishment, antiscience, and antiprogress zealots. The establishment's cavalier attitude toward such legitimate concerns is exemplified by a 1991 proposal of the USDA to release Australian wasps to combat a grasshopper problem on the Wyoming range. Fortunately a group of college students at the University of Wyoming (*Science*, 15 November 1991, p. 245) effectively derailed this absurd government proposal by demanding a full environmental impact assessment of releasing an alien species. It is inconceivable that a government agency should even consider such an absurd proposal, yet there is a minority faction in agribusiness that has no regard for or understanding of ecological farming and no qualms over the potential hazards of releasing nonindigenous or genetically engineered life forms into the environment.

Modern agriculture operates so close to the edge of disaster that the U.S. Department of Agriculture Research Service (ARS) spent $23.5 million in 1991 on biological control programs. Exotic foreign weeds, such as the field star thistle, and bugs such as the Russian wheat aphid, accidentally imported in contaminated agricultural produce and seeds, along with indigenous (nonimported) agricultural pests such as grasshoppers and medflies, are a serious and costly problem. According to ARS biological control specialist Jack Coulson, "annual savings from our successful biocontrol programs (for both native and intro-duced pests), based primarily on the costs of pesticides no longer required, total over $155 million per year."[3]

The rationale of using high-risk "biological controls"— such as releasing Australian wasps to control grasshoppers —as an alternative to more costly pesticides is the same rationale that the biotech industry is employing to justify the use of various genetically engineered products as alternatives to chemical pesticides. It is now a matter of public record (and of considerable chagrin for biotech-nology entrepreneurs) that agricultural pests are rapidly becoming resistant to *Bacillus thuringiensis* (*Bt*), a bacterium that produces a toxin poisonous to many insects. This bacterium has been sprayed on crops and genetically spliced into a variety of crops such as cotton and tomato. To help avoid insects becoming resistant, scientists advise farmers to mix transgenic seeds with normal seeds, a proposal that biotech companies don't like. According to Ann Gibbons:

> For these plans to work, they have to be put into action, and the experts meeting in Washington last week learned that they can't expect much help from the government on that score. Ann Lindsay, director of the Environmental Protection Agency's registration division in the office of pesticide pro-grams, disappointed workshop scientists by telling them that the agency had no plans to regulate the use of new biopesticides. This means that the only hope for convincing farmers to use *Bt* sparingly is to persuade them that it's in their interest to forsake some short-term income (by allowing insects to damage some of their crops) to make sure they have *Bt* as long as possible. But whether struggling farmers will be any more farsighted

---

[3] J. D. Beard, "Bug Detectives Crack the Tough Cases." *Science* 254 (1991): 1581.

than the average U.S. corporate executive remains
to be seen.[4]

The plethora of "biological immigrants"—exotic plant
and insect pests that are a threat to agriculture and are
responsible for millions of dollars of crop damage and loss
in the United States every year—should serve as a warning
to those who see no problems in releasing new genetically
engineered life forms into the environment. Some of these
biological immigrants include the blue water hyacinth
(from South America) and hydrilla (from Southeast Asia)
that are now clogging Florida's waterways; the Eurasian
carp and other deliberately introduced foreign fish species
that have decimated indigenous fish species across the
United States; and the Middle East sweet potato whitefly
(which was estimated to cause $200 million worth of
crop loss in California in 1991). Other ecologically harmful exotics include Eurasian Kentucky bluegrass, the
Africanized honeybee, Chinese kudzu, the Asian tiger
mosquito, European purple loosestrife, the European star-
ling, and the zebra mussel. Although many exotics have
been deliberately introduced, many come in accidentally
in agricultural produce and imported plants and seeds.
An estimated 10 percent of established immigrants have
major adverse ecological consequences, but there is a
new urgency because of the "homogenization of the world"
via import and export of agricultural commodities and
deliberate introduction of new plant and animal species
(including exotic African "game" being raised on Texas
ranches).[5]

4   "Moths Take the Field Against Biopesticide." *Science* 254(1991):
    646.
5   For further details see E. Culotta, "Biological Immigrants Under
    Fire." *Science* 254(1991): 1444–1447.

Conclusion

There are several interrelated dimensions to fully evaluating the costs and consequences, risks and benefits, of new developments in science, technology, and industry, especially in genetic engineering biotechnology and the patenting of both processes and products. These dimensions are ethical and spiritual, moral and religious, legal and political, social and economic, and environmental and cultural. Generally these dimensions of concern, constraint, and direction have been ignored by policymakers and even seen as obstacles to economic growth and industrial expansion. As a consequence, the gap between private (corporate) and public interests has widened—as evidenced by the rise of a global industrial biotechnocracy. The costs and consequences, risks and benefits, of this new world order need to be rigorously evaluated. Such concern should not be misjudged as antiscience or antiprogress sentiment. Rather, it should be recognized that only with greater involvement of an informed public in the policy-making process will advances in science and technology—and in biotechnology in particular—be likely to serve the public good and help enhance the quality of life and environment alike. Current attempts by the U.S. government to deregulate the biotechnology industry and by the EEC's Commission on Biotechnology to eliminate socioeconomic considerations in the licensing of new genetically engineered animal drugs support the conclusion that the direction being taken by the biotechnocracy of the industrialized world is neither prudent nor appropriate.

CORPORATE RESPONSIBILITY IN BIOTECHNOLOGY

Really we create nothing.
We merely plagiarize nature.
—JEAN BAITALLON

The idea that governments should now restrict man's
freedom, in the interest of animals, shows every
sign of getting its way.
—*THE ECONOMIST*, 16 November 1991, p. 21.

*T*hrough genetic engineering biotechnology we are on the threshold of ever-increasing control over the genetic blueprints of all living beings, from microbes and plants to mice and men. But who is responsible for this newfound power over the genes of life?

As some see it, we are playing God; others see the acquisition of biotechnology as a natural consequence of evolution. Regardless, we have the power to improve existing varieties of food plants in terms of disease and drought resistance and nutrient value; to develop new bioindustrial products from genetically altered plants and microorganisms, such as growth hormone, insulin, and monoclonal antibodies. Bacteria that can digest crude oil and remove industrial and agrichemical pollutants from contaminated water could be widely used to help restore environmental quality. Biotechnology is also revolutionizing medicine. It holds great promise not only in the treatment and prevention of genetic (inherited) diseases such as sickle-cell anemia and hemophilia but also in the diagnosis and treatment of a host of diseases from AIDS to cancer.

However, as with the application of new scientific knowledge to any commercial industry, there are risks as well as benefits to be considered. Not since we acquired power over the atom has our planetary dominion had such fundamental implications in terms of the future and integrity of creation. Consequently, corporate responsibility is a critical issue.

The gulf between enlightened self-interest (and the public good) and corporate interest must be bridged. This can be accomplished only by what I call ethical objectivity and holistic thinking. Ethical objectivity entails a rigorous assessment of the ethical, moral, and legal ramifications of biotechnology independent of vested interests. A holistic approach necessitates a broader view than a simplistic, linear, cost (risk)/benefit analysis. For example, the social, environmental/ecological, and long-term economic and cultural consequences of biotechnology must be carefully considered on a case-by-case basis.

Corporate responsibility clearly entails more than simplistic short-term cost/benefit assessments, public-acceptance surveys, and market projections. Likewise, it is too simplistic to dismiss all public concerns over biotechnology as irrational. There are legitimate concerns that the biotechnology industry should not ignore or paint over with public-relations campaigns designed to promote the public's unconditional acceptance of genetic engineering.

And it is surely irresponsible for the biotechnology industry to rely on the data and conclusions of its own scientists and technical consultants. The vested interests of such persons, if not their own conceptual biases, call into question their ability to provide a truly reliable and objective assessment of the complex ramifications of the many new applications of biotechnology.

What is needed is the establishment of interdisciplinary bioethics advisory boards to guide the biotechnology industry along the safe middle road of appropriate application and profitable innovation. Advisory boards for all the major fields in which biotechnology is being applied should include representatives from private industry, academia (including, where appropriate, experts in such fields as theology, philosophy, ethnobotany, and ecology), and public-interest groups (such as national farming, consumer, humane, and environmental organizations). Each of these three categories should have equal representation. Such an arrangement would help protect the interests of both the public and private sectors and reduce the direct and indirect costs of having federal and state agencies oversee the biotechnology industry, which many public-interest groups and the General Accounting Office have shown to be ineffectual and inefficient.

Various federal agencies are in the process of devising a host of regulations and guidelines for the biotechnology industry. A federal bioethics council would help integrate their activities and facilitate cooperation with both the private and public sectors. This council should have the same kind of interdisciplinary composition as proposed above for bioethics advisory boards, as well as having a representative from each federal agency that is now involved in regulating the industry.

With this kind of structure, the process of approval of new biotechnologies and products would be greatly accelerated, potential risks would be significantly minimized, and costly lawsuits could be avoided. (But this does not mean that companies creating potentially high-risk products—such as new vaccines and bacterial pesticides—should not be heavily insured.)

In view of the global scope of biotechnology, international standards and regulations need to be established and rigorously upheld. This would help overcome unfair competition between individual companies and remove the temptation of multinational corporations and their subsidiaries to engage in unethical and illegal activities, such as releasing genetically engineered organisms that are neither licensed nor approved in the United States into the environment elsewhere. Various drugs (such as depo-provera) and pesticides (such as DDT) banned in the United States have been sold and distributed in third-world countries. A repetition of this scenario by the highly competitive biotechnology industry could be disastrous.

It is surely unethical and unwise for corporations to invest in public-relations campaigns to coerce public opinion to favor biotechnology rather than to inform and convert. The industry should not feed false hope to the public through the media. To even speculate that genetic engineering could mean the end of all human suffering from famine, pestilences, and plagues is going too far. It reflects a naive "genetic determinism" that ignores the fact that these problems are caused by many factors that have little to do with genes.

Just as a sound economy can only be built upon a sound ecology, so corporate growth and profits are best built upon public trust. In the 1980s, more than one scientist and biotechnology company violated that trust by deliberately releasing genetically engineered microorganisms into the environment before obtaining governmental approval (see Chapter 3).

Biotechnologists should also think twice before inserting human genes into other animals (which has already been done with a variety of species) and in creating chimeras such as goat-headed sheep. Such things strike an uneasy chord with an informed public that is already sensitized by media coverage of illegal releases and demonstrations against the approved release of genetically engineered microorganisms.

Such public demonstrations of concern and fear should not be dismissed as some neo-Luddite antiscience and technology minority movement. The concerns and fears about the misapplication of biotechnology are very real. Any corporation or corporate executive that does not respect and share these concerns and fears is not being rational. It is the height of irrationality and denial to put short-term profits before evaluating the possibility of causing serious if not irreparable damage to the environment.

There are also some ironies to face. For years the public has listened to the petrochemical and pharmaceutical companies' denials that chemical pesticides are harmful to life—including human life, born and unborn. Now some companies claim that the new generation of genetically engineered pesticides will be more effective and much safer than the standard chemical pesticides, thus admitting, at long last, that the latter aren't so safe after all.

Another irony is in the belief that progress will be faster if there is a greater collaboration between universities and biotechnology corporations that provide research funds. On the contrary, progress will most likely be retarded, because science progresses best when there is both competition and international cooperation, especially in basic research. Corporate interests and trade secrets will stifle any cooperation or sharing of information. Scientific freedom and creativity are also at stake, since some of the most significant breakthroughs come not through intense, mission-oriented research, but through free thinking and basic, rather than applied, research.

The obvious trend toward monopolistic control of seed stock on a worldwide basis, followed rapidly by control of domestic animal germ plasm and gene lines, may not be in the best interest of a diversified and ecologically sound agriculture. This monopolistic trend, already endorsed by patent protection for new varieties of seeds and genetic engineering techniques, will accelerate further if the U.S. Patent and Trademark Office's ruling—that genetically engineered animals can be patented—is not rescinded or at least subjected to a lengthy moratorium for reasons discussed earlier.

This new technology will never go away. Nor will its foreseeable and potentially harmful residual consequences, unless it is applied with a new sensitivity and respect for life and its interconnected processes and for society and its interdependent parts. Where was the sensitivity and respect of those companies that poisoned the entire food chain with pesticides and harmed many lives, especially those of the sick and the unborn, with inadequately tested and often superfluous and ineffectual drugs? It is naive to blame such problems simply on greed —the profit motive. This same lack of sensitivity and respect for life and for the future integrity of the environment may well be the nemesis of biotechnology. Or perhaps these problems stem in part from a misguided altruism that places human interests in the wrong perspective, such that the human-centered worldview is ultimately self-destructive.

Dr. Barry Commoner has raised another important issue concerning corporate responsibility.[1] The temptation

---

[1] "Bringing Up Biotechnology," *Science for the People* 19(1987): 9–12.

to produce new genetically engineered products that do not maximize the benefits to society is very real. He points out that products that help prevent disease are generally less profitable than those used to treat disease, and the latter may be less profitable than those used for diagnostic and screening purposes. Yet disease prevention is the number one priority in terms of the public's interest. But the profits to biotechnology are not sufficiently attractive in such areas, even though the human need is great. For example, to spend R&D dollars on developing a malaria vaccine or a cure for sickle-cell anemia, as altruistic as it may seem, is not a good investment, since the markets would be low income and small volume, respectively. In such instances, government support of private research might be appropriate.

Commoner cites the Catholic bishops' letter on the economy that stated, "No one can own capital resources completely or control their use without regard for others and society as a whole." This echoes a recent encyclical by the Pope that emphasized that since capital is created by workers and the rest of society, society ought to have the right to determine what is done with that capital.

This societal right, in relation to biotechnology, is further strengthened by the fact that public funds were used to support the initial basic and applied research in biotechnology, which in turn was made possible by the wealth of knowledge acquired by the past efforts of publicly supported scientists all over the world. This accumulated knowledge (which is another form of capital) is a cultural heritage and common public property. In sum, biotechnology, as with any technological innovation, should not be used primarily to maximize private profits or the gross national product, but rather should satisfy people's basic needs, their right to creative participation and democratic control, as well as ecological soundness.

Commoner emphasizes that:

The biotechnology industry has repeated, step for step, what has happened in the petrochemical industry. If it is allowed to go much further, like the petrochemical industry, it will become invulnerable to control. Now at this early stage, we need to control what the biotechnology industry produces. A major test is to show that the product is so socially important as to outweigh the inherent risks. This means that a trivial product—with no social value at all—is worth no risk at all and simply should not be produced.

In conclusion, the issue of corporate responsibility in the field of genetic engineering biotechnology is neither simple nor trivial. The future and integrity of creation, as well as the well-being of the people and of all life on this planet, now hang in the balance. And a balance is needed to determine how this new technology is to be applied. The benefits to society and to corporations must be equivalent, and there can be no adverse impact on the environment or on the health and well-being of those who dwell therein. Although "environmentally neutral" biotechnology is a valid goal, a better one would be an environmentally and socially enhancing biotechnology.

This is an attainable goal. It is not wishful thinking. Mistakes will be made, but such risks can be minimized and this new technology can be applied creatively and profitably if there is corporate responsibility. And this responsibility for the integrity and future of creation is indeed as great as the power we now have over the gene and over life itself.

CHAPTER 9

BIOTECHNOLOGY, ETHICS, AND HUMAN EVOLUTION

We are rapidly approaching the point in our biological evolution when we will have one final choice—suicide or adoration.
—TEILHARD DE CHARDIN

Charles Darwin, in *The Descent of Man*, saw the human species as a descendant of the apes. His view of our relationship to the animal kingdom certainly upset the linear, hierarchical view of the Church, which embraced Aristotle's vision of a great chain of being with *Homo sapiens* at the top. It has been said that Darwin used to write on his hand every day "not superior," to remind himself to remind others that his theory of evolution should not be interpreted as an extension of Aristotle's view of human superiority, which the patriarchal church hierarchy of the times found so self-affirming.

Yet even today, many of those who accept the theory of evolution believe that it implies a kind of hierarchical progression toward some goal of perfection or personal or collective attainment. Scientific and technological advances are seen as evidence of this evolutionary process and as progress toward the utopian perfection of self and society. The desecration of the natural world and escalating extinction of plant and animal species are dismissed variously as the unavoidable price of progress, as a divinely ordained or biologically inevitable transformation of the natural world into a human industrialized one, or as evidence that only the fittest survive. Indeed the belief in survival of the fittest has given a quasi-scientific sanction to the destruction of the natural world and competitive extinction of nonhuman species. This erroneous belief ignores the evidence of interspecies cooperation, interdependence, and coevolution in nature. And it has made us unfit to survive on this planet, for as the late ecologist Gregory Bateson was fond of saying, "the organism that destroys its environment, destroys itself."

It is ironic that this misperception of nature has led to the widespread acceptance of competitive individualism and cutthroat competition within and between the industrial empire-states of the modern world as natural and therefore acceptable. Industrial competitiveness and expansion (like a parasitic, pioneer, or predator species displacing others) is condoned as natural and healthy by those who may be called latter-day social Darwinists.

Social Darwinism is the philosophy behind the rising biotechnocracy. It sees competition, survival of the fittest, and evolution toward some future utopian state of technological perfection as the highest values. Because of its

distorted perception of nature's "laws" and dynamics, social
Darwinism is blind to its own pathology. No nation-state
(or individual therein) can be self-sustaining if it lives like
a predator or parasite upon others, including the other
nation-states of kindred species like the deer, bear, wolf,
eagle, dolphin, and whale.

When I use the term *evolution* I do not mean the
progressive, instrumental "perfection" of humankind, but
rather the development, collectively, of the species, in
terms of such qualities as wisdom, humility, compassion,
empathy, and self-realization. The notion of the perfectibility of *Homo sapiens* and of nature alike not only distracts
us from being responsibly engaged in the here and now, it
also places greater emphasis on our "becoming" rather
than upon the quality of our existential being. Hence,
social Darwinism with its mythos of evolutionary 'progress'
is future oriented.

This cognitive distortion is masked by the belief that
through science and technology humanity and society can
and will be perfected and nature "improved" in the
process. The *means* by which perfection may be attained
are of no ethical significance to social Darwinists: Since
perfection is the supreme goal and the highest value, then
even evil means are considered morally acceptable if they
promise such laudable ends. An extreme example of this
process in action under the guise of altruism (the good of
society) and scientific validity (genetic improvement) was
the genocide policy of the Nazis. It is analogous to the
"species-cide" policy of contemporary biological fascism,
under which plant and animal species with no foreseeable
utility are exterminated and their natural ecosystems
converted into profitable bioindustrial systems.

The misapplication of biotechnology can be expected to
continue as long as this anthropocentric goal of perfecting
the rest of creation for the good of society prevails. And if
this attitude toward the perfectibility of the human species

prevails, we will see history repeat itself as we enter an age of eugenics. The historical amnesia of many biotechnologists is cause for concern. Contrast the following views: Dr. W. French Anderson contends that, "Enhancement genetic engineering presents significant, and troubling ethical concerns. Except where this type of therapy can be justified on the grounds of preventive medicine, enhancement engineering should not be performed."[1] Leroy Walters, an ethicist who heads Georgetown University's Center for Bioethics and serves on the National Institutes of Health (NIH) Recombinant DNA Committee, has stated publicly that he sees no ethical problem in germ-line therapy, even when the purpose is not curing disease but genetic enhancement, as of memory.[2] Although germ-line therapy in humans is still a distant dream, the Boston-based Committee for Responsible Genetics petitioned the NIH in 1986 to declare formally that it would entertain no proposals "that could alter germ-line cells" because this would be "tantamount to experimentation on future generations with no possibility of informed consent. It would also set up a direct path to programs of eugenics."

The philosophy of social Darwinism regards such theological and ethical concepts as the sanctity of being and the integrity of creation as immaterial, thus immeasurable and therefore nonexistent. To speak of the inherent nature and sanctity of any sentient being, of the integrity of species and of sentience per se, is to commit the scientific heresy of nonrational, subjective, "mystical" thinking. Instrumental rationalism and the acquisition of objective knowledge-as-power are the hallmarks of scientism that give a quasi-religious sanction to the technocratic worldview of today in the name of truth, progress, human evolution, and material benefit. This emerging worldview

[1]   *Journal of Medicine and Philosophy* 10(1985): 275–291.
[2]   *San Francisco Chronicle*, 29 September, 1987, p. A7.

that sees no ethical issue in the patenting of life or in genetically redesigning the Earth's creation should concern all reasonable people who feel and acknowledge the connections between humane planetary stewardship, peace, justice, and the integrity and future of creation. As a consequence of ontological reductionism, the perceived inherent value of living things is supplanted by the objective and instrumental valuation of their genes purely in terms of human utility.

There is a strangeness factor or element of the absurd in the minds of some reductionists. For instance, in 1976 Oxford University professor Richard Dawkins wrote a book entitled *The Selfish Gene* (Oxford University Press), in which he proposed that our bodies are simply the means that genes have evolved to ensure their immortality.

However, many "establishment" scientists endorse scientism as the key to human evolution. This belief has become the equivalent of a religious creed to biotechnologists and others who embrace the technocratic worldview of a bioindustrial utopia to come. Philosopher Mary Midgley, in her book *Evolution as a Religion: Strange Hopes and Stranger Fears* (New York: Methuen, 1985), writes, "Evolution seemed to endorse egoism and thereby capitalism. Despite protests from both scientists and philosophers, people still find this interpretation irresistible." Furthermore, since it is believed that genetic engineering will be used to "improve" upon the human race, we have the mythic goal of the superman reemerging. As Dr. Midgley emphasizes, Nietzsche sanctified this mythic goal in his philosophy, which was built in part upon the Thomistic theology that all of creation is not for the veneration of God, but for man's use as a god over nature. In precisely this vein, the well-known Harvard University biologist E. O. Wilson, in his book *On Human Nature* (Harvard University Press, 1978) proclaims:

The time has come to ask: Does a way exist to divert the power of religion into the service of the great new enterprise that lays bare the sources of that power? . . . Make no mistake about the power of scientific materialism. It presents the human mind with an alternative mythology that until now has always, point for point in zones of conflict, defeated traditional religion.

According to Midgley, the central difficulty with the technocratic vision of utopia and faith in techniques like genetic engineering to stimulate the progressive evolution of *Homo sapiens* "is that this story is arbitrarily human-centered and that its view of humanity is at present arbitrarily intellect-centered." She goes on to state:

Its human-centeredness distorts both evolutionary theory and our attitude to the natural world. By what right, and in what sense, can we consider ourselves as the directional pointer and aim-bearer of the whole evolutionary process? Does this mean what is often taken for granted today in controversy about the treatment of plants and animals, that all other organisms exist only as means to our ends? Kant and other philosophers have said this, many people believe it, yet it remains extremely obscure. The idea that things are *there* for some external purpose seems to need a theological context, and this view did of course grow out of one. But that context will not subjugate everything to man. Certainly Judaeo-Christian thinking made the human race much more central than many other religions do, but it still considered man to be God's steward. Divine aims were always paramount, and God had created all his creatures for his own purposes, not for man's. Non-human beings count in this picture as having their own special value. Redwoods and pythons, frogs, moles and albatrosses are not failed humans or early try-outs for humans or tools put there to advance human development.

The danger of ontological reductionism apparent in the technocratic attitude toward living systems and sentient beings (that are nothing more than gene machines, protein factories, or assemblies of cells) is that the worldview (and *the kind of science and technology that arise therefrom*) is conceptually flawed. The renowned physicist David Bohm makes the following poignant observation:

> Molecular biologists have discovered that in the growth and reproduction of cells, certain laws that can be given a mechanical form of description are satisfied (especially those having to do with DNA, RNA, the synthesis of proteins). From this, most of them have gone on to the conclusion that ultimately *all* aspects and sides of life will be explained in mechanical terms. But on what basis can this be said? . . . It should be recalled that at the end of the nineteenth century, physicists widely believed that classical physics gave the general outlines of a complete mechanical explanation of the universe. Since then, relativity and quantum mechanics have overturned such notions altogether. . . . Classical physics was swept aside and overturned. . . . Is it not likely that modern molecular biology will sooner or later undergo a similar fate? . . .
>
> The notion that present lines of thinking will continue to be validated indefinitely by experiment is just another article of faith, similar to that of the nineteenth-century physicists. . . . Is there not a kind of "hubris" that seems rather often to penetrate the very fabric of scientific thought, and to capture the minds of scientists, whenever any particular scientific theory has been successful for some period of time? This takes the form of a fervently held belief that what has been discovered

will continue to work indefinitely, ultimately to cover the whole of reality.[3]

The ontological reductionism espoused by the scientism of genetic engineering biotechnology is undoubtedly the greatest threat to our species' spiritual, if not physical, survival—as well as to the integrity and future of creation. For many reasons—economic, ideological, and political— genetic engineering to benefit primarily human ends cannot be stopped. The momentum of scientific advances has propelled humanity into a new dimension of power over the Earth's creation. If we are to maximize the benefits and minimize the risks, we cannot allow this momentum, nor the explosive growth of the biotechnology industry, to continue in a moral and ethical vacuum that precludes both awareness of and responsibility for the consequences.

It is crucial for us all—theologians, religious leaders policy makers and educators, especially—to recognize that no human venture, including biotechnology, can proceed in a "leaderless" fashion without moral and ethical guidelines. What should these be? Reverence for all life and the view that all sentient beings are worthy of moral consideration are the antithesis of those self-serving values of technocratic thought and perception. What duties do we have to assure the future and integrity of the rest of creation? What is the best way to exercise our power of dominion, and especially our newly acquired power over the genes of life? From a theological perspective, what is our proper role in the Earth process and created order? How can appropriately applied biotechnology enhance this

---

[3] "On the Subjectivity and Objectivity of Knowledge," in *Beyond Chance and Necessity*, J. Lewis, ed. (London: Garnstone Press, 1974).

role, if indeed it is to preserve the natural world in all
its beauty, integrity, and diversity? What are our moral
obligations toward other sentient beings under our
dominion, and what ethical principles are there, if any, to
help assure the proper application of genetic engineering
biotechnology? Is there a theological basis for the moral
imperative to protect wildlife and wildlands from the
potentially harmful consequences of biotechnology and to
protect animals subjected to genetic engineering from
harm? Is the possibility of animal suffering the only issue,
or do individual animals and species have a sovereign right
to genetic integrity, making it a violation of the sanctity of
sentient life to introduce human and other species' genes
into their genomes? Even if they do not suffer, the genetic
integrity of so-called transgenic animals is harmed to a
degree and rate that are unprecedented under either wild
conditions of natural selection or domestic conditions of
traditional selective breeding.

These and other questions were raised at a consultation
I convened of religious leaders and representatives of
major Christian denominations. They issued the following
statement less than two weeks after the U.S. Patent and
Trademark Office granted the first patent for a genetically
engineered mouse to Harvard University:

On Ethics and Theology
We affirm that humanity and all of nature live
in a relationship of mutuality and interaction in
covenant with the Creator.
We recognize that the human species is not in
right relationship with the rest of creation; and
that our transgression lies in our continued abuse
of the creation and our desire to remake it in our
own image as a means of satisfying exclusively
human ends. Redemption includes not only personal salvation but also the restoration of the

natural world and establishment of a relationship that will protect the integrity of creation.

The ethical, environmental, socioeconomic and theological ramifications of genetic engineering and patenting of life are profound. They point to the probability that the integrity and future of creation will be placed in even greater jeopardy if our power over the genes of life is not exercised prudently and with reverence to help to restore the covenant: to heal the Earth and ourselves.[4]

There is as yet no consensus about how best to apply biotechnology, which is illustrative of a diversity of attitudes toward life and the natural environment. Daniel E. Koshland, Jr., writing in *Science* magazine (236[1987]: 1157), concludes:

Ecological and moral dilemmas created by these new technologies are appreciable and will require new ideas. Tampering with the mind is generally considered to be bad, but should genetic engineering to alleviate Alzheimer's disease be outlawed? To feed starving populations is desirable, but if new crops help add a billion people to a crowded globe, is that necessarily good? The powerful new methods are here. Applying them may well require the use of brain enhancers.

Biotechnology is not only a product of the evolution of human knowledge and ingenuity, it is also an evolutionary force in itself. It is forcing us to ask the kinds of questions Dr. Koshland raises, and by so doing, it becomes a catalyst for an entirely new way of thinking and acting. This technology is part of the evolving consciousness and power of *Homo sapiens*. It becomes, at the same time, a call to

---

[4] Consultation on "Respect for Life and the Environment: Moral and Theological Aspects of Genetic Engineering and Biotechnology," Airlie, VA, April 1988.

conscience. We are "choosers" as well as "doers," and the moral choices that are made in terms of how this new technology is applied will determine our own destiny and the fate of the biosphere. It may indeed be our last hope to restore the health and vitality of this poisoned, desecrated, and fast dying planet. Greed and arrogance on the one hand, and fear and desperation on the other, must not stand in the way of the appropriate and creative applications of biotechnology. There are those, on the one extreme, that would use biotechnology for short-term gain regardless of potentially harmful long-term consequences; on the other side are those that would oppose all genetic engineering because history has taught us how easily and destructively we abuse our powers of dominion. Biotechnology, seen as an evolutionary force in itself, as a kind of biofeedback upon the consciousness and conscience of humanity, could well mean the apotheosis and epiphany of *Homo sapiens*.

The sage advice of Mahatma Gandhi needs to be followed: "Civilization should not be the infinite multiplication of human wants. Science should not order human values. Technology should not order society. Civilization should be the deliberate limitation of wants to essentials that could be equally shared."

Some three hundred years ago at the beginning of the industrial revolution, John Donne wrote, "Nature is the common law by which God governs us." If we do not obey this common law and we continue to parasitize the Earth and exploit all that dwell therein, allowing scientific materialism to order human values and technology to order society, the fruits of biotechnology will be a bitter harvest indeed.

As Rousseau concluded, the voice of nature in us is our conscience. And our conscience is awakening like a survival instinct as we face the stark reality that we have built our present civilization at nature's expense. We are our own worst enemy, since in the process of seeking security through power, we are destroying the very things we seek to protect—the land, society, our cultural institutions, and those values that make us civilized and without which the demise of our culture and the natural world will be inevitable. The plight of desperate peoples—the poor, the hungry, the uneducated, and the unemployed—of endangered species and vanishing cultures, and the world-wide suffering of humans and animals under the tyranny of industrial-military control, are a call to conscience. The voice of nature in us may yet save us from ourselves.

Concerned ecologists of today urge that we learn to walk lightly on the Earth. Wherever we tread, we leave a mark that may take generations to heal, if ever. And these will be generations of suffering, be it from destroying the forests; silting up the rivers; creating deserts, acid rain, or poisoned seas; destroying the ozone layer; or adding to global warming.

Since the advent of fire and pyrotechnology, each generation has left a deeper and more permanent mark on the Earth. But no mark could ever be deeper or more permanent than changing the entire direction of the Earth's evolutionary process and the ecologically integrated communities of wild animals and plants, through genetic engineering, primarily to serve our own ends. My hope is that this new technology will instead be used to heal the mistakes of past generations and help us restore and preserve the planet's natural biodiversity, so that we will enjoy far better health and prosperity than we have for many generations.

This is not a utopian dream. It is simply a matter of choice and consequences. There are those who believe that humankind and nature are somehow separate and that we can prosper at nature's expense. But there are others—perhaps the last human beings left who have not forgotten their history or lost their natural wisdom—who know that we cannot. This does not mean that we should not exploit life in order to sustain our own. It is a matter of degree. Commercial exploitation of genetically engineered animals that has nothing to do with ecologically sound and sustainable agriculture and holistic preventive medicine is ethically and morally unacceptable. It is also economically unacceptable because this new technology is nonsustainable. It robs future generations of what few options they might have left to walk gently on the Earth and remove the carnage of an ignorant past.

Even if we cannot atone for the past, surely we can care for the future of creation and choose to live wisely. We need not abandon biotechnology as some evil or Promethean curse, but we must use it prudently, with respect, humility, and compassion. This is indeed as much a challenge to corporate America as it is to the rising global biotechnocracy. With the support of an informed and concerned public, and of their political representatives, the industrial, commercial, agricultural, and medical applications of genetic engineering could indeed serve the greater good, which includes not just the good of society but the integrity and future of all creation.

A P P E N D I X   A

Milestones in the History of Genetic Engineering and Biotechnology

1869   Breeding experiments with pea plants by Austrian monk Gregor Mendel reveal the mechanisms of heredity.

1944   O. T. Avery and co-workers at Rockefeller University discover that genes are made of DNA.

1953   British scientists Francis Crick and James Watson describe the double-helix structure of DNA.

1965   RNA is used to synthesize proteins in a test tube. Sequences of three nucleotides specify amino acids to be linked together to form proteins by the ribosome. The genetic code is broken.

1970   Hamilton Smith and Daniel Nathans of Johns Hopkins University discover a new class of restriction enzymes that act like scissors, cutting DNA strands in precise locations.

1973   Stanley Cohen of Stanford University and Herbert Boyer of the University of California at San Francisco insert ("splice") recombinant DNA into host bacteria that reproduce (clone) copies of the foreign DNA as they multiply. The technique is patented, and the age of genetic engineering begins.

1976   Genentech in San Francisco, California, becomes the first corporation formed to commercially develop the new technology.

1977   Frederick Sanger of the British Medical Research Council and Walter Gilbert of Harvard University independently discover techniques for the rapid sequencing, or reading, of the order of nucleotides in DNA molecules.

1980   U.S. Supreme Court rules 5 to 4 that an oil-eating bacterium developed by a General Electric Company researcher can be patented, overturning a policy of the Patent and Trademark Office.

1980   Animal production technologists at Ohio University in Athens transfer genes from other species into mice to create the first transgenic animals.

1986–   Government permits are given that allow the first
87   release of genetically engineered organisms into the environment: Biologics Corp. of Omaha, with a genetically engineered swine vaccine; Advanced Genetic Sciences Inc. of California, with a genetically altered bacterium ("Frost-ban").

1992   An estimated 10,000 different kinds of transgenic mice have been created. Pigs are being genetically altered to provide replacements for human "Xenografts". The Bush administration refuses to sign an international biodiversity protection agreement at the Earth Summit in Brazil to protect U.S. biotechnology interests.

A P P E N D I X   B

*Organizations that Supported
Legislation to Halt the Patenting
of Animals*

Animal Welfare Organizations

American Anti-Vivisection Society
American Fund for Alternatives to Animal Research
American Humane Association
American Society for the Prevention of Cruelty to Animals
Animal Defense Council, Inc.
Animal Legal Defense Fund, Inc.
Animal Protection Institute of America, California
Animal Welfare Institute
Association of Veterinarians for Animal Rights
Farm Animal Reform Movement
Friends of Animals, Norwalk, CT
Fund for Animals, New York
Humane Farming Association
Humane Society of the United States
International Association Against Painful Experiments on
    Animals
International Fund for Animals, Barnstable, MA
International Society for Animal Rights, Inc.

Jews for Animal Rights
Massachusetts Society for the Prevention of Cruelty to
    Animals
Michigan Humane Society
National Alliance for Animal Legislation
National Anti-Vivisection Society, Chicago and
    Washington
New England Anti-Vivisection Society
Women's SPCA of Pennsylvania

Agricultural Organizations

American Agriculture Movement
Center for Rural Affairs
International Alliance for Sustainable Agriculture
Land Institute
Land Stewardship Project
League of Rural Voters
National Farm Organization
National Farmers Union
National Save the Family Farm Coalition
Rural Advancement Fund
United Farmers Organization

Environmental and Public-Interest Organizations

Defenders of Wildlife (pending formal approval)
Elsa Wild Animal Appeal
Foundation on Economic Trends
Friends of the Earth
Medical Research Modernization Committee
Physicians Committee for Responsible Medicine
Psychologists for the Ethical Treatment of Animals
United States Public Interest Research Group

APPENDIX C

Resources
for Public Involvement

There are several resources for concerned citizens to learn more about developments in biotechnology and how it may affect their lives and the future of all life on this planet. The following regularly published newsletters deal with developments in biotechnology as they relate to human and environmental health and safety and to the economy and well-being of third-world countries:

- *The Ag Bioethics Forum*, an interdisciplinary newsletter in agricultural bioethics, The Ag Bioethics Forum, 115 Morrill Hall, Iowa State University, Ames, Iowa 50011.

- *Biotechnology and Development Monitor*, joint publication of the Directorate General, International Cooperation of the Ministry of Foreign Affairs, The Hague, and the University of Amsterdam, The Netherlands.

- *Genetic Engineering News*, a monthly industry magazine, c/o Harris Marketing Systems, 700 Mt. Prospect Avenue, Newark, NJ 07104.

- *GeneWatch*, a bulletin of the Council for Responsible Genetics, 19 Garden Street, Cambridge, MA 02138.

- *Global Pesticide Campaigner*, Pesticide Action Network, 965 Mission Street, San Francisco, CA 94103.

- *The Gene Exchange: A Public Voice on Genetic Engineering*, National Biotechnology Policy Center, National Wildlife Federation, 1400 16th Street, NW, Washington, DC 20036.

- *RAFI Communique*, Rural Advancement Fund International, P.O. Box 665, Pittsboro, NC 27312.

- *West Coast Biowatch*, 2700 H Street, Sacramento, CA 95816.

More technical, industry, and science-related publications include:

- *Genetic Engineering Letter*, 8750 Georgia Avenue, Suite 124, Silver Spring, MD 20910.

- *Biotechnology: The International Monthly for Industrial Biotechnology*, Nature Publishing Co., 65 Bleeker, New York, NY 10012-2467.

- *AgBiotechnology: Agricultural Research/Business News*, P.O. Box 7, Cedar Falls, IA 50613.

The Council for Responsible Genetics

The following publications are available from the Council for Responsible Genetics, 19 Garden Street, Cambridge, MA 02138; phone, (617) 868-0870; fax, (617) 864-5164.

- *CRG Position Paper on the Human Genome Initiative* ($3). This statement provides a critique of the project that has been heralded as providing a genetic explanation of who we are and how we function. In its position, the Council illuminates why claims of the project's usefulness are exaggerated from a scientific, social, and ethical perspective.

- *CRG Position Paper on Genetic Discrimination* ($3). This statement demonstrates how the expansion in the number and range of genetic tests has opened the door to one of the oldest forms of discrimination, the classification of people based on their genes, particularly in the areas of employment and insurance.

- *Position Paper on Large Scale Release of Genetically Engineered Microorganisms* ($3). As the products of biotechnology become part of our foods, medicines, and environment, serious questions arise about how to prevent the biotechnology industry from going down the same path as the chemical or nuclear industry. The Council calls for a moratorium on the large-scale commercial releases of genetically engineered microorganisms based on the regulatory system's inability to evaluate the risks of such releases.

- *Preventing a Biological Arms Race* ($19 + postage). CRG is pleased to announce this publication edited by Board Member Dr. Susan Wright. The book presents essential political, technical, and legal information about biological weapons and biological disarmament. Frank Barnaby, former Director of the Stockholm International Peace Research Institute, calls it "essential reading for researchers and all interested in armament and disarmament issues." MIT Press, 446 pages. Please add postage of $2.90 within the United States, $3.72 in Canada or Mexico, $8.29 in Europe, and $11.04 in Asia.

- *Biotechnology's Bitter Harvest: Herbicide-Tolerant Crops and the Threat to Sustainable Agriculture* ($5). This report, prepared by the Biotechnology Working Group, provides the most comprehensive analysis to date of the development of herbicide-tolerant crops by the biotechnology industry and the environmental and public health risks of increased use of harmful herbicides. A question-and-answer brochure accompanies this report.

Office of Technology Assessment Reports

The following OTA reports are available from the U.S. Government Printing Office, Superintendent of Documents, Dept. 33, Washington, DC 20402-9325, (202) 783-3238; or the National Technical Information Service, 5285 Port Royal Road, Springfield, VA 22161-0001, (703) 487-4650.

- *Ownership of Human Tissues and Cells—Special Report*, OTA-BA-337, March 1987; 176 p.

- *Public Perceptions of Biotechnology*, OTA-BP-BA-45, May 1987; 136 p.

- *Field-Testing Engineered Organisms: Genetic and Ecological Issues—Special Report*, OTA-BA-350, May 1988; 160 p.

- *U.S. Investment in Biotechnology—Special Report* (free summary available from OTA), OTA-BA-360, July 1988; 340 p.

- *Patenting Life—Special Report*, OTA-BA-370, April 1989; 204 p.

- *Commercial Biotechnology: An International Analysis*, OTA-BA-218, January 1984; 616 p.

- *Impacts of Applied Genetics: Micro-Organisms, Plants, and Animals*, OTA-HR-132, April 1981; 332 p.

- *Mapping Our Genes: Genome Projects—How Big? How Fast?* OTA-BA-373, April 1988; 232 p.

- *Biotechnology in a Global Economy*, OTA-BA-494, October 1991; 283 p.

Glossary

antibody. A protein that binds to an invading *antigen* in the body prior to the destruction of the antigen.

antigen. A foreign substance, usually a protein, that triggers the body's production of an antibody directed specifically to neutralize the antigen.

bacterium. A single-celled microscopic organism with a primitive nucleus.

chromosomes. The rod-like structures in the cells of the body carrying the genes, the number of which vary from species to species. Humans have a set of 46 chromosomes in every cell; each chromosome carries some 100 to 200,000 genes.

clone. One or more genetically identical organisms. Bacteria clone naturally. Frogs and calves have been artificially cloned.

DNA. Deoxyribonucleic acid, a complex chain-like nucleotide shaped in the form of a double helix. A gene is a segment of DNA. The particular sequencing of chemicals in a segment of DNA determines what information is stored in a particular gene.

enzymes. Some of these chemicals are used like scissors to cut a desired gene segment of DNA. These are called restriction enzymes. Other enzymes called ligases are used to splice the isolated gene into the DNA of a living organism, such as a bacterium. The bacterium multiplies by dividing to produce billions of identical *clones* that carry the new gene that programs them to produce insulin or other desired proteins.

genes. The smallest segment of DNA containing a hereditary message. Each gene is a basic unit of heredity, passing traits on from one generation to the next.

genetic code. The sequence of nucleotides that determines protein synthesis in cells.

hormone. A substance secreted by one type of cell that carries a signal to influence the activity of another type of cell.

monoclonal antibodies. Identical antibodies that have been cloned from a single source and targeted for a specific *antigen*.

mutation. A random change in genetic material that can change a host of functions and may be lethal.

nucleotide. The building block of the DNA molecule consisting of an organic base, a phosphate, and a sugar.

plasmid. A self-replicating circular molecule of DNA found in bacteria and carrying two or more genes.

protein. A molecule of linked amino acids. Proteins serve principally as structural parts of an organism and as biochemical catalysts.

recombinant DNA. A new combination of genes spliced together on a single piece of DNA. Also refers to a group of techniques for cutting and splicing.

RNA. Ribonucleic acid—the nucleic acid that transmits and translates DNA's genetic instructions, or the genetic material itself in some organisms.

Selected Bibliography

BRUNNER, E. *Bovine Somatotropin: A Product in Search of a Market*. London: London Food Commisson, 1988.

BUSCH, L., LACEY, W. B., BURKHARDT, J., and LACEY, L. R. *Plants, Power, and Profit*. Oxford: Basil Blackwell, 1991.

COOPER, IVER P. *Biotechnology and the Law*. New York: Clark Boardman Co., 1989.

DAYAN, A. D., CAMPBELL, P. N., and JUKES, T. H., eds. *Hazards in Biotechnology: Real or Imaginary*. Proceedings of the Biological Council's symposium in London, 14–15 December 1987. New York: Elsevier, 1989.

DOYLE, JACK. *Altered Harvest*. New York: Viking, 1985.

ELKINGTON, J. *The Gene Factory: Inside the Biotechnology Business*. London: Carroll & Graf, 1985.

FCCSET. *Biotechnology for the 21st Century*. Report from the Committee on Life Sciences and Health, Federal Coordinating Council for Science, Engineering & Technology. Washington, DC: U.S. Government Printing Office, 1992.

FIKSEL, J., and COVELLO, V. T., eds. *Safety Assurance for Environmental Introductions of Genetically-Engineered Organisms*. New York: Springer-Verlag, 1988.

GENDEL, STEVEN M., KLINE, DAVID A., WARREN, D. MICHAEL, and YATES, FAYE, eds. *Agricultural Bioethics*. Ames: Iowa State University Press, 1990.

GOODMAN, D., SORJ, B., and WILKINSON, J. *From Farming to Biotechnology: A Theory of Agro-Industrial Development*. London: Basil Blackwell, 1987.

GUSSOW, Joan Dye. *Chicken Little, Tomato Sauce And Agriculture: Who Will Produce Tomorrow's Food?* New York: The Bootstrap Press, 1991.

HACKING, ANDREW J. *Economic Aspects of Biotechnology*. Cambridge: Cambridge University Press, 1987.

HASSEBROOK, CHUCK, and HEGYES, GABRIEL. *Choices for the Heartland: Alternative Directions in Biotechnology & Implications for Family Farming, Rural Communities & the Environment*. Ames: Iowa State University Press, 1989.

HOBBELINK, HENK. *Biotechnology and The Future of World Agriculture*. London: Zed Books. 1991.

HURNIK, FRANK, and LEHMAN, HUGH, eds. *Ethics and Agricultural Biotechnology: Opposing Viewpoints*. Vol 4, No. 2. 1991 *Journal of Agricultural and Environmental Ethics*. Ontario: University of Guelph.

JACOBSSON, S., JAMISON, A., and ROTHMAN, H., eds. *The Biotechnological Challenge*. Cambridge: Cambridge University Press, 1986.

JONES, STEPHANIE. *The Biotechnologists*. New York: Macmillan, 1992.

JUMA, CALESTOUS. *The Gene Hunters: Biotechnology and the Scramble for Seeds*. Princeton: Princeton University Press, 1989.

KENNEY, MARTIN. *Biotechnology: The University-Industrial Complex*. New Haven, CT: Yale University Press, 1986.

KLOPPENBURG, JACK R. Jr. *First the Seed: The Political Economy of Plant Biotechnology.* Cambridge: Cambridge University Press, 1988.

KRIMSKY, S. *Biotechnics & Society. The Rise of Industrial Genetics.* New York: Praeger, 1991.

LAPPÉ, MARC. *Broken Code: The Exploitation of DNA.* San Francisco: Sierra Club Books, 1985.

MOLNAR, JOSEPH J., and KINNUCAN, HENRY, eds. *Biotechnology and the New Agricultural Revolution.* Boulder, CO: Westview Press, 1989.

NATIONAL AGRICULTURAL BIOTECHNOLOGY COUNCIL REPORTS. 1. *Biotechnology and Sustainable Agriculture: Policy Alternatives;* 2. *Agricultural Biotechnology, Food Safety and Nutritional Quality for the Consumer.* 3. *Agricultural Biotechnology at the Crossroads: Biological, Social and Institutional Concerns.* Ithaca, NY: Boyce Thompson Institute for Plant Research, Cornell University, annual series.

NATIONAL RESEARCH COUNCIL. *Field Testing Genetically Modified Organisms.* Washington, DC: National Academy Press, 1989.

PERSLEY, G. J. *Beyond Mendel's Garden: Biotechnology in the Service of World Agriculture.* Washington, DC: CAB International and the World Bank, 1990.

STRAUSS, HARLEE S. *Biotechnology Regulations: Environmental Release Compendium.* Washington, DC: OMEC International, 1987.

SUZUKI, DAVID, and KNUDTSON, PETER. *Genethics.* Cambridge, MA: Harvard University Press, 1989.

TEITELMAN, ROBERT. *Gene Dreams.* New York: Basic Books, 1989.

WHEALE, P., and MCNALLY, R., eds. *The BioRevolution: Cornucopia or Pandora's Box.* London: Pluto Press, 1990.

WOODMAN, WILLIAM F., SHELLY, MACK C. II, and REICHEL, BRIN J. *Biotechnology and the Research Enterprise. A Guide to the Literature.* Ames: Iowa State University Press, 1989.

WORLD BANK. *Agricultural Biotechnology: The Next "Green" Revolution?* Technical paper no. 133. Washington, DC: World Bank, 1990.

WRIGHT, SUSAN, ed. *Preventing a Biological Arms Race.* Cambridge, MA: MIT Press, 1990.

YOXEN, EDWARD. *The Gene Business.* New York: Harper & Row, 1984.

INDEX